ESSAYS ON LIFE ITSELF

Complexity in Ecological Systems Series

Essays on
LIFE ITSELF

ROBERT ROSEN

COLUMBIA UNIVERSITY PRESS

NEW YORK

Columbia University Press
Publishers Since 1893
New York   Chichester, West Sussex

Copyright © 2000 by Columbia University Press
All rights reserved

Library of Congress Cataloging-in-Publication Data
Rosen, Robert, 1934–1998
    Essays on life itself   /   Robert Rosen.
        p.   cm.   —   (Complexity in ecological systems series)
    Includes bibliographical references.
    ISBN 0–231–10510–X (hardcover   :   alk. paper).   —   ISBN 0–231–10511–8
    (pbk.   :   alk. paper)
        1. Life (Biology)—Philosophy.   I.   Title.   II.   Series.
QH501.R655   1999
570'.1—dc21   99–24034

∞

Casebound editions of Columbia University Press books are printed on permanent
and durable acid-free paper.
Printed in the United States of America
c 10 9 8 7 6 5 4 3 2 1
p 10 9 8 7 6 5 4 3 2 1

On behalf of my father, Robert Rosen, who passed away December 30, 1998, I would like to dedicate this book to the responsible pursuit of knowledge. The main curiosity driving my father's life and scientific career revolved around the question, "What makes living things alive?" In the course of trying to answer that question, he discovered many things, he created a whole new area of science, and he traveled all over the world. He also found his answers, many of which are contained and discussed in these pages. The answers were rarely where he expected them to be, however, and one of the discoveries he made was that not everyone in science wants to "follow the problem" where it leads or hear the truth. He also discovered that knowledge can be abused or perverted and he occasionally decided not to publish all he had learned.

If the information and ideas in this volume help others in their own quest for knowledge, my father would join me in the hope that readers will exercise the same concern and sense of responsibility he always felt in putting these ideas to use.

May you enjoy the journey as much as he did.

—JUDITH ROSEN

Contents

ॐ

Contents

PREFACE

❧

This volume is a collection of essays, intended primarily to enlarge upon a number of points that were touched upon in *Life Itself.* I believe they are of independent interest and importance, but I felt the ideas could not be pursued in that place, since they would detract from the main line of the argument to which *Life Itself* was devoted.

Thus this volume should be considered a supplement to the original volume. It is not the projected second volume, which deals with ontogenetics rather than with epistemology, although some chapters herein touch on ideas to be developed therein.

The essays presented here were mainly written after *Life Itself* was published. They were prepared for special occasions: meetings, seminars, conferences, and workshops. Some of them have appeared in print before, as contributions to the *Proceedings* volumes arising from these occasions. They were never intended for journals: the subject matter and the treatment required put them outside the scope of any journal known to me that would reach the appropriate audiences. I have on occasion submitted one or another of them to general journals such as *Science,* but the manuscripts were not even refereed.

Nevertheless, I feel there is sufficient interest in this material, and in the issues with which they deal, to justify bringing them together in one place, and in this way. Indeed, I have been surprised and gratified by the general reaction to *Life Itself.* I have received more correspondence relating to this volume than any other publication of mine, and from a broader spectrum of readers, from orthodox molecular biologists to software developers, linguists, and social scientists. For one reason or another, these correspondents expressed their covert uneasiness with previously presented paradigms; it was a very practical uneasiness, a feeling that their problems were not actually being addressed from those directions. They saw in *Life Itself* a language in which more satisfactory, and more practical, alternatives to current orthodoxies could be expressed. That is exactly what I had hoped.

This correspondence, in fact, reminded me of my fairly extensive travels in eastern Europe years ago. The orthodoxy there, at that time, was Dialectical Materialism, which also promised the solution to everything. Everyone avowed it: It was mandatory to do so. But no one really believed it.

ESSAYS ON LIFE ITSELF

Part I

∞

ON BIOLOGY AND PHYSICS

T HE CHAPTERS in part I are essentially the text of a brief talk presented at a workshop on "Limits to Scientific Knowability," held at the Santa Fe Institute (SFI) in 1994, May 24 to 26. As described to me, the workshop was intended to explore the impacts (if any) of the famous Gödel Incompleteness results in mathematics upon the sciences. The workshop's tone was to be informal and exploratory, aimed at determining whether a more extensive effort by the Institute along these dimensions was warranted.

Accordingly, the workshop consisted primarily of roundtable discussion, with no formal papers delivered. However, the organizers requested a few participants to deliver brief position statements about the impacts of noncomputability results on their field. I was asked to do this for the field of biology. The following is, as best I recollect it, a reconstruction of what I said on this occasion. I have said it all before, but rarely so succinctly, rarely with such a feeling that I was saying exactly what I wanted to say. I include it here as a general introduction.

I was impressed by the general air of affability that pervaded this workshop. I had been rather critical of the SFI's activities, and most particularly of their programs in "complexity" and in "artificial life." My misgivings had arisen from a conviction that the future of graduate education and innovative basic science in this country rests on the development of private research institutes such as the SFI, and that limitless harm would be done if such a heavily promoted endeavor were to embark down unfruitful scientific paths. I do not want to see that happen—the situation is already fragile. However, my experience in this workshop reassured me that the SFI is basically healthy, that it has much to contribute, and that it should be supported by all those concerned with the principles involved.

To me, the basic question in biology, to which all others are subsidiary or collateral, is the one put most succinctly by the physicist Erwin Schrödinger: What is life?

Any question becomes unanswerable if we do not permit ourselves a universe large enough to deal with the question. $Ax = B$ is generally unsolvable in a universe of positive integers. Likewise, generic angles become untrisectable, cubes unduplicatable, and so on, in a universe limited by rulers and compasses.

I claim that the Gödelian noncomputability results are a symptom, arising within mathematics itself, indicating that we are trying to solve problems in too limited a universe of discourse. The limits in question are imposed in mathematics by an excess of "rigor," and in science by cognate limitations of "objectivity" and "context independence." In both cases, our universes are limited, not by the demands of problems that need to be solved but by extraneous standards of rigor. The result, in both cases, is a mind-set of reductionism, of looking only downward toward subsystems, and never upward and outward.

In science, for instance, it seems patently obvious that, whatever living organisms are, they are material systems, special cases drawn from a larger, more generic class of nonliving inorganic ones. The game is thus to *reduce*, to express their novel properties in terms of those of inorganic subsystems, merely subject to a list of additional conditions and restrictions. Indeed, one manifestation of this claim to the objectivity of reduction is that one must never, ever, claim to learn anything new about matter from a study of organisms. This is but one of the many forms of the protean Central Dogma (Judson 1979), expressed here as a limitation on material nature itself.

Despite the profound differences between those material systems that are alive and those that are not, these differences have never been expressible in the form of a list—an explicit set of conditions that formally demarcate those material systems that are organisms from those that are not. Without such a list, Schrödinger's question, and biology itself, become unanswerable at best, meaningless at worst. So we must probe more deeply into what the quest for such a list actually connotes.

No such list means there is no algorithm, no decision procedure, whereby we can find organisms in a presumably larger universe of inorganic systems. It has of course never been *demonstrated* that there is no such list. But no one has ever found one. I take seriously the possibility that there is no list, no algorithm, no decision procedure, that finds us the organisms in a presumptively larger universe of inorganic systems. This possibility is already a kind of noncomputability assertion, one that asserts that the world of lists and algorithms is too small to deal with the problem, too nongeneric.

Indeed, the absence of lists or algorithms is a generally recurring theme in science and mathematics, one that reveals the nongenericity of the world of algorithms itself, a world too unstable (in a technical sense) to solve the real problems. This was the upshot of the Gödel results from the very beginning.

It helps to recall the mathematical situation that Gödel inherited. It was a world still reeling from the discovery of non-Euclidean geometries almost a century earlier, geometries without number that were just as consistent as Euclid was. It was a world reeling from paradoxes within Cantorian set theory. There had to be something to blame for all of this; something to be expunged, to make everything right again; something not rigorous enough, which had to be found and eradicated.

Bertrand Russell, among others, argued that the fault lay in "impredicative" definitions and vicious circles, and he developed an elaborate and murky "theory of types" to replace them with predicative but equivalent counterparts. This was taken yet further by Hilbert and his school of formalists; they argued that rigor lay entirely in syntax, and that the difficulties at the foundations of mathematics arose entirely from unextruded, semantic residues of meaning. For them, a mathematical term (e.g., *triangle*) was not to be allowed any vestige of meaning; rather, there were to be formal production rules for manipulating *triangle* from one proposition to another. This drastic extrusion of semantics constituted true rigor; mathematics itself would be suspect as long as there was any vestige of meaning or semantics left in it. Hilbert sought this kind of formalization of all of mathematics, the reduction of mathematics to algorithms or lists.

It was this program that Gödel's results killed. Briefly, these results mean that a constructive universe, finitely generated, consisting of pure syntax, is too poor to do mathematics in. They mean that semantics and impredicativities and meanings are essential to mathematics; they cannot be replaced by more syntactic rules and more lists or algorithms. They mean that mathematical systems are generically unformalizable; hence it is the formalizable ones that are the rare special cases, and not the other way around. They mean that identifying rigor with formalizability makes most of mathematics unreachable.

I argue that biology teaches us that the same is true about the material world. Roughly, that contemporary physics is to biology as Number Theory is to a formalization of it. Rather than an organism being just a standard material system plus a list of special conditions, an organism is a repository of meanings and impredicativities; it is more generic than an inorganic system rather than less. If this is so, then the Schrödinger question, and indeed biology itself, is not exclusively, or even mainly, an empirical science; empirics is to it as accounting is to Number Theory.

If this is so, then organisms possess noncomputable, unformalizable models. Such systems are what I call *complex*. The world of these systems is much larger and more generic than the simple world we inherit from reductionism.

The main lesson from all this is that computability, in any sense, is not itself a law of either nature or mathematics. The noncomputability results, of which Gödel's was perhaps the first and most celebrated, are indicative of the troubles that arise when we try to make it such.

CHAPTER I

ℭ

The Schrödinger Question, *What Is Life?*
Fifty-Five Years Later

Erwin Schrödinger's essay *What Is Life?*, which first appeared in print in 1944, was based on a series of public lectures delivered the preceding year in Dublin. Much has happened, both in biology and in physics, during the half century since then. Hence, it might be appropriate to reappraise the status of Schrödinger's question, from a contemporary perspective, at least as I see it today. This I shall attempt herein.

I wonder how many people actually read this essay nowadays. I know I have great difficulty in getting my students to read anything more than five years old, their approximate threshold separating contemporary from antiquarian, relevant from irrelevant. Of course, in the first decade or two of its existence, as H. F. Judson (1979) says, "everybody read Schrödinger," and its impacts were wide indeed.

The very fact that everybody read Schrödinger is itself unusual, for his essay was a frank excursion into theoretical biology, and hence into something that most experimental biologists declare monumentally uninteresting to them. Actually, I believe it was mostly read for reassurance. And, at least if it is read superficially and selectively, the essay appears to provide that in abundance—it is today regarded as an utterly benign pillar of current orthodoxy.

But that is an illusion, an artifact of how Schrödinger's exposition is crafted. Its true messages, subtly understated as they are, are heterodox in the extreme and always were. There is no reassurance in them; indeed, they are quite incompatible with the dogmas of today. By the stringent standard raised in the Schrödinger title question, following these dogmas has actually made it harder, rather than easier, to provide an adequate answer.

What Is Life?

Let us begin with the very question with which Schrödinger entitled his essay. Plainly, this is what he thought biology was *about,* its primary object of study. He thought that this "life" was exemplified by, or manifested in, specific organisms, but that at root, biology was not about *them*—it concerned rather whatever it was about these particular material systems that distinguished them, and their behaviors, from inert matter.

The very form of the question connotes that Schrödinger believed that "life" is in itself a legitimate object of scientific scrutiny. It connotes a noun, not merely an adjective, just as, say, rigidity, or turbulence, or (as we shall see later) openness does. Such properties are exemplified in the properties or behaviors of individual systems, but these are only *specimens;* the concepts themselves clearly have a far wider currency, not limited to any explicit list of such specimens. Indeed, we can ask a Schrödinger-type question, What is X? about any of them.

I daresay that, expressed in such terms, the Schrödinger question would be dismissed out of hand by today's dogmatists as, at best, meaningless; at worst, simply fatuous. It seems absurd in principle to partition a living organism, say a hippopotamus, or a chrysanthemum, or a paramecium, into a part that is its "life," and another part that is "everything else," and even worse to claim that the "life" part is essentially the same from one such organism to another, while only the "everything else" will vary. In this view, it is simply outrageous to regard expressions like "hippopotamus life" or "chrysanthemum life" to be meaningful at all, let alone equivalent to the usual expressions "living hippopotamus" and "living chrysanthemum." Yet it is precisely this interchange of noun and adjective that is tacit in Schrödinger's question.

This approach represents a turnabout that experimentalists do not like. On the one hand, they are perfectly willing to believe (quite deeply, in fact) in some notion of *surrogacy,* which allows them to extrapolate their data to specimens unobserved; to believe, say, that *their* membrane's properties are characteristic of membranes in general, or that the data from their rat can be extrapolated ad libitum to other species (Rosen 1983; see my *Anticipatory Systems* for fuller discussion). On the other hand, they find it most disquieting when their systems are treated as the surrogatees, and especially to be told something about *their* membrane by someone who has not looked at their membrane, but rather at what they regard as a physicomathematical "abstraction." When pressed, experimentalists tend to devolve the notions of surrogacy they accept on *evolution;* surrogates "evolve" from each other, and, hence, what does not evolve cannot be a surrogate. One cannot have the issue both ways, and that is one of the primary Schrödinger unorthodoxies, tacit in the very question itself.

A typical empiricist (not just a biologist) will say that the Schrödinger question is a throwback to Platonic Idealism and hence completely outside the pale of science. The question itself can thus be entertained only in some vague metaphoric sense, regarded only as a *façon de parler,* and not taken seriously. On the other hand, Schrödinger gives no indication that he intends only such metaphoric imagery; I think (and his own subsequent arguments unmistakably indicate) that, to the contrary, he was perfectly serious. And Schrödinger knew, if anyone did, the difference between Platonism and science.

Schrödinger and "New Physics"

Erwin Schrödinger was one of the outstanding theoretical physicists of our century, perhaps of any century. He was a past master at all kinds of propagation phenomena, of statistical mechanics and thermodynamics, and of almost every other facet of his field. Moreover, he viewed physics itself as the ultimate science of material nature, including of course those material systems we call organisms. Yet one of the striking features of his essay is the constantly iterated apologies he makes, both for his physics and for himself personally. While repeatedly proclaiming the "universality" of contemporary physics, he equally repeatedly points out (quite rightly) the utter failure of its laws to say anything significant about the biosphere and what is in it.

What he was trying to say was stated a little later, perhaps even more vividly, by Albert Einstein. In a letter to Leo Szilard, Einstein said, "One can best feel in dealing with living things *how primitive physics still is*" (Clark 1972; *emphasis added*).

Schrödinger (and Einstein) were not just being modest; they were pointing to a conundrum about contemporary physics itself, and about its relation to life. Schrödinger's answer to this conundrum was simple, and explicit, and repeated over and over in his essay. And it epitomized the heterodoxy I have alluded to before. Namely, Schrödinger concluded that organisms were repositories of what he called *new physics*. We shall turn a little later to his gentle hints and allusions regarding what that new physics would comprise.

Consider, by contrast, the words of Jacques Monod (1971), writing some three decades after the appearance of Schrödinger's essay:

> Biology is *marginal* because—the living world constituting but a tiny and very "special" part of the universe—it does not seem likely that the study of living things will ever uncover general laws applicable outside the biosphere. (*emphasis added*)

With these words Monod opens his book *Chance and Necessity*, which sets out the orthodox position. This idea of the "marginality" of biology, expressed as a denial of the possibility of learning anything new about matter (i.e., about physics) by studying organisms, is in fact the very cornerstone of his entire development.

Monod did not dare to attack Schrödinger personally, but he freely condemned anyone else who suggested there might be "new physics" wrapped up in organism, or in life, in the harshest possible way; he called them vitalists, outside the pale of science. Sydney Brenner, another postulant of contemporary orthodoxy, was even blunter, dismissing the possibility of a new physics as "this nonsense."

But Schrödinger, within his own lifetime, had seen, and participated in, the creation of more new physics than had occurred in almost the entire previous history of the subject. It did not frighten him; on the contrary, he found such possibilities thrilling and exhilarating; it was what he did physics for. Somehow, it is only the biologists it terrifies.

There is one more historical circumstance that should perhaps be mentioned here. Namely, biological thoughts were lurking very close to the surface in the cradles of the New Quantum Theory in the 1920s. Niels Bohr himself was always profoundly conscious of them. He had in fact grown up in an atmosphere of biology; his father (for whom the familiar Bohr effect, involving the cooperativity of binding of oxygen to hemoglobin, was named) was an eminent physiologist. Many of Bohr's philosophical writings, particularly those dealing with complementarity, are awash in biological currents (Pais 1991). In general, the creators of the New Quantum Theory believed they had at last penetrated the innermost secrets of all matter. I have been told, by numerous participants and observers of these developments, of the pervasive expectation that the "secrets of life" would imminently tumble forth as corollaries of this work.

That, of course, is not what happened. And indeed, Schrödinger's ideas about the new physics to be learned from organisms lie in quite a different direction, which we shall get to presently.

Genotypes and Phenotypes

We have seen in the preceding chapters just how radical and unorthodox Schrödinger's essay is, first in simply posing the question What is life? and second in tying its answer to new physics. Both are rejected, indeed condemned, by current dogmas, which cannot survive either of them. How, then, could this essay possibly have been read for reassurance by the orthodox?

The answer, as I have hinted, lies in the way the essay is crafted. Viewed superficially, it looks primarily like an exposition of an earlier paper by Schrödinger's younger colleague, Max Delbrück.[1] Delbrück, a student during the yeasty days in which the New Quantum Theory was being created, was deeply impressed by the ambiences I have sketched here. Indeed, he turned to biology precisely because he was looking for the new physics Schrödinger talked about, but he missed it. Delbrück's paper, on which Schrödinger dwelt at such length in his essay, argued that the "Mendelian gene" had to be a molecule (but see later section, "Order from Order").

Today, of course, this identification is so utterly commonplace that no one even thinks about it any more—a deeply reassuring bastion of reductionism. But it is in fact much more complicated than it looks, biologically and, above all, physically. As we shall see shortly, identifications require two different processes, and Delbrück argued only one. It was Schrödinger's attempt to go the other way, the hard way, roughly to deal with the question, When is a molecule a Mendelian gene? that led him to his new physics, and thence to the very question, What is life?

At this point, it is convenient to pause to review the original notion of the Mendelian gene itself, a notion intimately tied to the genotype-phenotype dualism.

Phenotypes, of course, are what we can see directly about organisms.

They are what behave, what have tangible, material properties that we can measure and compare and experiment with. Gregor Mendel (*Life Itself,* section 11C) originally conceived the idea of trying to account for the similarities, and the differences, between the phenotypes of parents and offspring in a systematic way.

Mendel was, at heart, a good Newtonian. Newton's Laws[2] in mechanics say roughly that if *behaviors* are differing, then some *force* is acting. Indeed, one recognizes a force by the way it changes a behavior, and that is how one measures that force. In these terms, Mendel's great innovation was to conceive of phenotypes as *forced behaviors,* and to think of underlying "hereditary factors" (later called genes) as forcers of these phenotypes. In a more philosophical parlance, his hereditary factors constituted a new causal category for phenotypes and their behaviors; he was answering questions of the form, Why these phenotypic characters? with answers of the form, Because these hereditary factors. Mendel proceeded to measure the forcings of phenotype by genotype, by selecting a particular phenotype (the wild type) as a standard and comparing it to phenotypes differing from it in only one allele, as we would now say.

Exactly the same kind of thing was then going on elsewhere in biology. For instance, Robert Koch was also comparing phenotypes and their behaviors; in his case, the characters were "healthy" (his analog of wild type) and "diseased." The differences between them, the symptoms or syndromes marking the discrepancy between the former and the latter, were also regarded as forced, and the forcers were called germs. This constituted the "germ theory" of disease.

To anticipate somewhat, we can see that any such genotype-phenotype dualism is allied to the Newtonian dualism between states (or phases) and forces. The former are what behave, the latter are what make them behave. In a still earlier Aristotelian language, the states or phases represent material causation of behavior; the forces are an amalgam of formal and efficient causation. In biology, the phenotypes are the states and behaviors, the genotypes or germs are identified as forces which drive them.

On the other hand, it is all too easy to simply posit forces to account for the tangible changes of behavior that we can see directly. Critics of science have always pointed out that there is indeed something ad hoc, even ineluctably circular, in all this—to define a force in terms of observed behavior, and then turn around and explain the behavior in terms of that posited force. Indeed, even many scientists regard the unbridled invention of such forces as the entire province of theory and dismiss it accordingly, out of hand, as something unfalsifiable by observation of behavior alone. Worst of all, perhaps, such a picture generally requires going outside a system, to a larger system, to account for behaviors inside it; this does not sit well with canons of reductionism, nor with presumptions of objectivity or context independence in which scientists like to believe. Finally, of course, we should not forget fiascoes such as phlogiston, the epicycles, and the luminiferous ether, among many others, which were all characterized in precisely such a fashion.

For all these reasons, then, many people doubted the reality of the Mendelian genes. Indeed, for similar reasons, many eminently respectable physicists doubted the reality of atoms until well into the present century (Pais 1982).

It is precisely at this point that the argument of Delbrück, which Schrödinger develops in such detail in his essay, enters the picture. For it proposes an identification of the functional Mendelian gene, defined entirely as a forcer of phenotype, with something more tangible, something with properties of its own, defined independently—a *molecule*. It proposes, as we shall see, a way to realize a force in terms of something more tangible that is generating it. But, as we shall now see, this involves a new, and perhaps worse, dualism of its own.

On Inertia and Gravitation

What we are driving toward is the duality between how a given material system changes its own behavior in *response* to a force, and how that same system can *generate* forces that change the behavior of other systems. It is precisely this duality that Schrödinger was addressing in the context of "Mendelian genes" and "molecules," and the mode of forcing of phenotype by genotype. A relation between these two entirely different ways of characterizing a material system is essential if we are to remove the circularities inherent in either alone.

To fix ideas, let us consider the sardonic words of Ambrose Bierce, written in 1911 in his *Devil's Dictionary*, regarding one of the most deeply entrenched pillars of classical physics:

GRAVITATION, n. The tendency of all bodies to approach one another, with a strength proportioned to the quantity of matter they contain—the quantity of matter they contain being ascertained by the strength of their tendency to approach one another. This is a lovely and edifying illustration of how science, having made A the proof of B, makes B the proof of A.

This, of course, is hardly fair. In fact, there are two quite different quantities of matter involved, embodied in two distinct parameters. One of them is called *inertial mass:* it pertains to how a material particle *responds* to forces imposed on it. The other is called *gravitational mass:* it pertains rather to how the particle *generates* a force on other particles. From the beginning, Newton treated them quite differently, requiring separate Laws for each aspect.

In this case, there is a close relation between the *values* of these two different parameters. In fact, they turn out to be numerically equal. This is a most peculiar fact, one that was viewed by Einstein not merely as a happy coincidence but rather as one of the deepest things in all of physics. It led him to his Principle of Equivalence between inertia and gravitation, and this in turn provided an essential cornerstone of General Relativity.

We cannot hope for identical relations between inertial and gravitational aspects of a system, such as are found in the very special realms of particle mechanics. Yet, in a sense, this is precisely what Schrödinger's essay is about. Delbrück, as we have seen, was seeking to literally reify a forcing (the Mendelian gene), something "gravitational," by clothing it in something with "inertia"—by *realizing* it as a molecule. Schrödinger, on the other hand, understood that this was not nearly enough, that we must also be able to go the other way and determine the forcings manifested by something characterized "inertially." In more direct language, just as we hope to realize a force by a thing, we must also, perhaps more importantly, be able to realize a thing by a force. It was in this latter connection that Schrödinger put forward the most familiar parts of his essay: the "aperiodic solid," the "principle of order from order," and the "feeding on negative entropy." And as suggested earlier, it was precisely here that he was looking for the new physics. We shall get to all this shortly.

Before doing so, however, we must look more closely at what this peculiar dualism between the inertial and the gravitational aspects of a material system actually connotes.

Newton himself was never much interested in understanding what a force *was;* he boasted that he never even asked this question. That was what he meant when he said, "Hypothesis non fingo." He was entirely interested in descriptions of system behaviors, which were rooted in a canonical state space or phase space belonging to the system. Whatever force "really" was, it was enough for Newton that it manifested itself as a *function* of phase, i.e., a function of something already inside the system. And that is true, even when the force itself is coming from *outside*.

This, it must be carefully noted, is quite different from *realizing* such a force with an inertia of its own, generally quite unrelated to the states or phases of the system being forced. This latter is what Schrödinger and Delbrück were talking about, in the context of the Mendelian gene, as a forcer of phenotype. As Newton himself did not care much about such realization problems, neither did the "old physics" that continues to bear his personality. Indeed, this is perhaps the primary reason that Schrödinger, who increasingly saw "life" as wrapped up precisely with such realization problems, found himself talking about new physics. It is the tension between these two pictures of force that will, one way or another, dominate the remainder of our discussion.

A central role was played in the original Newtonian picture by the *parameters* he introduced, exemplified by "inertial mass" and "gravitational mass." Roughly, these serve to couple states or phases (i.e., whatever is behaving) to forces. In mechanics, these parameters are independent of both phases and forces, independent of the behaviors they modulate. Indeed, there is nothing in the universe that can change them or touch them in any way. Stated another way, these parameters are the quintessence of objectivity, independent of any context whatever.

Further, if we are given a Newtonian particle, and we ask what kind, or "species," of particle it is, the answer lies neither in any particular *behavior* it manifests under the influence of one or another force impressed on it, nor in the states or phases that do the behaving, but rather precisely in those parameter values—its masses. They are what determine the particle's identity, and in this sense *they are its genome*. The particular behaviors the particle may manifest (i.e., how its phases or states are changing when a force is imposed on it) are accordingly only *phenotypes*. Nor does this identity reside in the behaviors of other systems, forced by it.

In causal language, these parameters constitute *formal cause* of the system's behaviors or phenotypes (the states themselves are their material causes, the forces are efficient causes).

Thus there is a form of the phenotype-genotype dualism arising already here, where *genome* (in the sense of species-determining or identity-determining) is associated with *formal causes* of behaviors or phenotypes. It arises here as a consequence of the dualism mentioned earlier, between the states or phases of a system and the forces that are making it behave. If these last remarks are put into the context of the realization problems that Schrödinger and, to a much lesser extent, Delbrück were addressing, it becomes apparent that the situation is not quite so straightforward as current dogmas would indicate. We will return to these matters shortly.

"Order from Order"

I will now digress from conceptual matters and look briefly at Schrödinger's essay into the realization problems I discussed earlier. In general, he was concerned with turning inertia into gravitation, a thing into a force, a molecule into a Mendelian gene. This is perhaps the most radical part of Schrödinger's argument, which ironically is today perceived as an epitome of orthodoxy.

Delbrück had argued that the Mendelian gene, as a forcer of phenotype, must be inertially realized as a molecule. The argument was as follows: Whatever these genes are, in material terms, they must be small. But small things are, by that very fact, generally vulnerable to thermal noise. Genes, however, must be stable to (thermal) noise. Molecules are small and stable to thermal noise. Ergo, genes must be molecules. Not a very cogent argument, perhaps, but the conclusion was satisfying in many ways; it had the advantage of being *anschaulich*, or visualizable. Actually, Delbrück's arguments argue only for *constraints*, and not just holonomic, Tinkertoy ones like rigidity; the same arguments are just as consistent with, for example, two molecules per "gene," or three molecules, or N molecules, or even a fractional part of a molecule.

Schrödinger was one of the first to tacitly identify such constraints with the concept of order. Historically, the term *order* did not enter the lexicon of physics until the latter part of the nineteenth century, and then only through an identification of its negation, *disorder*, with the thermodynamic notion of entropy. That is, something was ordered if it was not disordered, just as something is nonlinear if it is not linear.

As I discussed in *Life Itself* (see section 4F), constraints in mechanics are identical relations among state or phase variables and their rates of change. If configurational variables alone are involved, the corresponding constraint is called *holonomic*. Rigidity is a holonomic constraint. The identical relations comprising the constraint allow us to express some of the state variables as functions of the others, so that not all the values of the state variables may be freely chosen. Thus, for example, a normal chunk of rigid bulk matter, which from a classical microscopic viewpoint may contain 10^{30} particles, and hence three times that number of configurational variables, can be completely described by only six. Such heavily constrained systems are often referred to nowadays as *synergetic*.[3] H. Haken (1977) calls the independently choosable ones controls, and the remaining ones slaved. We might note, in passing, that traditional bifurcation theory[4] is the mathematics of breaking constraints; its classic problems, like the buckling of beams and other failures of mechanical structures, involve precisely the breaking of rigid constraints as a function of changing parameters associated with impressed *forcings*.

Nonholonomic constraints, which involve both configuration variables and their rates of change, have received much less study, mainly because they are not mathematically tidy. However, they are of the essence to our present discussion, as we shall see.

The language of constraints as manifestations of order can be made compatible with the language of entropy coming from thermodynamics, but the two are by no means equivalent. Schrödinger took great pains to distinguish them, associating the latter with the old physics, embodied in what he called "order from disorder," marking a transition to equilibrium in a closed system. But by speaking of order in terms of constraints, he opened a door to radically new possibilities.

Schrödinger viewed phenotypes, and their behaviors, as *orderly*. At the very least, the behaviors they manifest, and the rates at which these behaviors unfold, are highly constrained. In these terms, the constraints involved in that orderliness are inherently nonholonomic, viewed from the standpoint of phenotype alone.

Delbrück had argued that a Mendelian gene (as a forcer of phenotype) was, in material (inertial) terms, a molecule, mainly on the grounds that molecules were rigid. Thus whatever order there is in a molecule entirely resides in its constraints. But these, in turn, are holonomic. As Schrödinger so clearly perceived, the real problem was to somehow move this *holonomic* order, characteristic of a molecule, into the *nonholonomic* order manifested by a phenotype (which is not a molecule). In the more general terms used in the preceding section, the problem is to realize an inertial, structural, holonomic thing in terms of a force exerted on a dynamic, nonholonomic thing.

This was the genesis of Schrödinger's conception of order from order, or, more precisely, large-scale, nonholonomic, phenotypic order being forced by small-scale, rigid, holonomic, molecular order. It was this kind of situation for which Schrödinger found no precedent in the old physics. This was why, in his eyes, organisms resisted the old physics so mightily.

Schrödinger expressed the holonomic order he perceived at the genetic end in the form of the aperiodic solid. In other words, not just *any* holonomic or rigid structure could inertially realize a Mendelian gene, but only certain ones, which both specialized and generalized conventional molecules in different ways. Nowadays, it is axiomatic to simply identify "aperiodic solid" with "copolymer," and indeed, with DNA or RNA, and the constraints embodying the holonomic order with "sequence." But this changing of names, even if it is justified (and I claim it is not), does not even begin to address the realization problem, the transduction of genomic inertia into gravitation that Schrödinger was talking about.

Schrödinger was perhaps the first to talk about this transduction in a cryptographic language, to express the relation between holonomic order in genome, and nonholonomic order in phenotype, as constituting a *code*. This view was seized upon by another physicist, George Gamow,[5] a decade later; after contemplating the then-new Watson-Crick structure for DNA, he proposed a way to use DNA as a template, for moving its holonomically constrained "order" up to another holonomically constrained but much less rigid inertial thing, protein. This is a very far cry from the code that Schrödinger was talking about; it is at best only an incremental syntactic step. The next big one would be to solve the protein-folding problem (see *Life Itself,* section 11F), something over which the old physics claims absolute authority. After three decades of fruitless, frustrating, and costly failures, the field is just beginning to move again. Ironically, this is being done by postulating that protein folding is a forced rather than a spontaneous process, and by trying to realize these putative forcers in inertial terms. Thus in a sense the Mendelian experience is being replayed in a microcosm. But this is another story.

In addition to the principle of order from order that Schrödinger introduced to get from genotype to phenotype, and the aperiodic solid that he viewed as constituting the genetic end of the process, and the idea of a cryptographic relation between holonomic constraints in genotype and the nonholonomic ones characterizing phenotype, Schrödinger introduced one more essential feature: the idea of *feeding* (on "negative entropy," he said, but for our purposes it does not matter what we call the food). This was not just a gratuitous observation on his part. He was saying that, for the entire process of order from order to work at all, the system exhibiting it *has to be open* in some crucial sense. In the next section, we shall look at this basic conclusion in more detail.

Molecular biologists, in particular, found reassurance in Schrödinger's essay, mainly because of his use of innocent-sounding terms in familiar contexts. However, whatever this essay may offer, it is not reassurance.

The "Open System"

Thus Schrödinger envisioned two entirely different ways in which biological phenotypes, considered as material systems, are open. On the one hand, they are open to forcings, embodied tacitly in the Mendelian genes. On the other hand, they are also open to what they feed on, what they "metabolize." The former involves the effects of something *on* phenotype; the latter involves the effects of phenotype on something else (specifically, on "metabolites" residing in the environment). Schrödinger was tacitly suggesting a profound connection between these two types of openness—namely, that a system open in the first sense must also be open in the second. Stated another way, the entire process of order from order that he envisioned, and indeed the entire Mendelian process that it represented, cannot work in a (thermodynamically) closed system at all.

Such thermodynamically open systems accordingly can be considered "phenotypes without genotypes." They are the kinds of things that Mendelian genes can force. So this is a good place to start, especially since, as we shall see, it is already full of new physics, even without any explicit genome to force it. To anticipate somewhat, we will be driving toward a new perspective on Schrödinger's inverse question, When can a molecule be a Mendelian gene? in terms of another question of the form, When can a thermodynamically open system admit Mendelian forcings?

The history of ideas pertaining to open systems is in itself interesting and merits a short statement.[6] The impetus to study them and their properties came entirely from biology, not at all from physics. Thinkers in the latter field preferred to rest content with closed, isolated, conservative systems and their equilibria, and to blithely assign their properties a universal validity.

The first person to challenge this, to my knowledge, was Ludwig von Bertalanffy[7] in the late 1920s. Ironically, he was attempting to combat the frank vitalism of the embryologist Hans Driesch, particularly in regard to "equifinal" embryological or developmental processes. Bertalanffy showed that these phenomena, which so puzzled Driesch, were understandable once we gave up the strictures of thermodynamic closure and replaced the concept of equilibrium by the far more general notion of steady state (*fliessgleichgewicht*) or the still more general types of attractors that can exist in open systems.[8]

Bertalanffy was a person whom Jacques Monod loathed, and whom he (among many others) castigated as a "holist." By their very nature, open systems require going outside a system, going from a smaller system to a larger one to understand its behaviors. Stated another way, openness means that even a complete understanding of internal parts or subsystems cannot, of itself, account for what happens when a system is open. This flies in the face of the "analysis," or reductionism, that Monod identified with "objective science." But this is another story.

In the late 1930s, Nicolas Rashevsky (see *Life Itself,* section 5B) discovered some of the things that can happen in a specific class of such open systems, presently termed reaction-diffusion systems. He showed explicitly how such systems could spontaneously establish concentration gradients in the large. This is, of course, the most elementary morphogenetic process, and at the same time it is absolutely forbidden in thermodynamically closed systems. It might be noted that another name for this process, in physiology, is active transport. Over a decade later, this process was rediscovered by Alan Turing[9] (1950) in a much simpler mathematical context than Rashevsky had used. A decade after that, the same phenomena were picturesquely characterized by Ilya Prigogine (a member of the Brussels School, writing from a base in irreversible thermodynamics), under the rubric of symmetry breaking. A huge literature on pattern generation, and self-organization in general, has arisen in the meantime, based on these ideas.

Bertalanffy himself was quite well aware of the revolution in physics that was entailed in his concept of the open system. Indeed, he said quite bluntly, "The theory of open systems has opened up an entirely *new field of physics*" (1952). Quite early in the game, Prigogine (1947) likewise said, "Thermodynamics is an admirable but *fragmentary* theory, and this fragmentary character originates from the fact that it is applicable only to states of equilibrium in closed systems. Therefore, it is necessary to establish a broader theory."

Even today there is no acceptable physics of open systems, which are not merely closed systems perturbed in a certain way (see, e.g., chapter 12). This is because closed systems are so degenerate, so nongeneric, that when opened, the resultant behavior depends on how they were opened much more than on what they were like when closed. This is true even for the classical theory of thermodynamics itself, and it is why this classical theory does not lend itself to expansion into a true physical theory of open systems. What passes for theory at this level is entirely phenomenological, and it is expressed in dynamic language, not thermodynamic. These facts are of direct and urgent concern to experimental analysis, particularly in biology, because the very first step in any analytic procedure is to open the system up still further, in a way that is itself not reversible. That is, roughly, why analysis and synthesis are not in general inverse processes (cf. later section, What About Artificial Life?).

In any case, Schrödinger himself *could* have known about these incipient revolutions in the old physics, tacit in systems that feed and metabolize. But he had fixed his attention entirely on molecules, and on biochemistry, and hence he missed a prime example of the very thing he was asserting, and which most biologists were even then denying, namely that organisms teach new lessons about matter in general.

Open systems thus constitute in themselves a profound and breathtaking generalization of old physics, based as it is on the assumption of excessively restrictive closure conditions, conservation laws, and similar nongeneric presumptions that simply do not hold for living things. Seen in this light, then, is it really biology that is, in Monod's words, "marginal," "a tiny and very special part of the universe," or is it rather the old physics? In 1944, Schrödinger suggested that it was the latter that might be the case. Today, fifty-five years later, that possibility continues to beckon and, indeed, with ever-increasing urgency.

The Forcing of Open Systems

The behaviors manifested in open systems, such as their capacity to generate and maintain stable spatial patterns, exemplify neither the classical thermodynamic notion of "order from disorder," as Schrödinger used the term, nor what he called "order from order." As I have said, open system behaviors look like phenotypes, but they are not forced, in any conventional sense, and certainly not in any Mendelian sense, even though they have "genomes" expressed in their parameters. Nevertheless, their behaviors can be stable without being rigid or in any sense holonomically constrained. Let us see what happens when we impose forcings on such a system and, especially, when we try to internalize those forcings.

The essence of an open system is, as we have seen, the necessity to invoke an "outside," or an environment, in order to understand what is going on "inside." That is, we must go to a larger system, and not to smaller ones, to account for what an open system is doing. That is why reductionism, or analysis, that only permits us to devolve system behavior upon subsystem behaviors, fails for open systems. And as we have seen, that is why there is so much new physics inherent in open systems. That fact, of course, does not make openness unphysical; it simply points up a discrepancy between the physics we presently know and the physics we need.

But there are many ways a system can be open. So far, I have discussed only thermodynamic openness, characterized by energetic and material fluxes through the system. These are characterized by corresponding sources and sinks generally residing outside the system itself, in its environment. Inherent in this view is the notion of the system exerting forces on its environment, acting as a pump and driving the flow from sources to sinks.

However, an open system in this thermodynamic sense can itself be forced: the environment can impress forces on the system. This is what we have called a gravitational effect, and it is in general a quite different kind of openness to environmental influence than the thermodynamic openness we have just been considering. System behavior under the influence of such impressed forces has always been the lifeblood of classical particle mechanics and also, in a somewhat modified form, of what is today roughly called Control Theory.[10]

If there is already much new physics in the free behaviors of open systems, we should not be surprised to find much more in their forced behaviors, especially since our intuitions about how material systems respond to impressed forces are generally drawn from very simple systems, indeed generally linear ones. One of these intuitions, embodied in such things as servomechanisms and homeostats, is that a forced system will generally end up tracking the forcing. If this is so, it is correct to say that the relation between such an impressed force and the resulting system behavior is ultimately a cryptographic one; the explicit relation between the two is embodied in the familiar *transfer function*[11] of the system. That is already suggestive, but it is very risky to simply extrapolate such ideas to open systems.

A system that is open in *any* sense is one whose behaviors depend on something outside the system itself, whereas in a closed system, there *is* no outside. Thus it has always been a tempting idea to internalize the external influences in some way, to get a bigger system that is closed and deal with that. Unfortunately, the genericity of openness forbids it; genericity in this sense means that openness is preserved under such perturbations (i.e., physical openness is structurally stable). Indeed, what we end up with in this fashion is generally a bigger open system, which is in some sense even more open than the one we started with. This is, in itself, an important observation, which among other things underlies the familiar notion of the side effect (see the discussion of side effects in my *Anticipatory Systems*). At any rate, what one typically ends up with after carrying out such a strategy is the entire universe, which is not very helpful.

In general, the unforced or free situation in any system is one in which every force in the system is an internal force. In the language introduced earlier, it is a situation in which every gravitational aspect in the system can be assigned to a corresponding inertial aspect of that system. On the other hand, if a force is impressed on such a system from outside, that force has no inertial correlate within the system; there is in some sense an excess of gravitation over available inertia, an "inertial defect," if you will.

Thus if we wish to try to internalize such a force, we must augment our original system with more inertia; in practice, that means adding more state variables and more parameters to the system, in such a way that the forced behavior of the original system is now free behavior of the larger system.

Now, the effect of any force is to modify a rate, compared with what that rate would be in an unforced or free situation. That is, a force shows up in the system as an acceleration or deceleration of some system behavior (i.e., it acts as a *catalyst*). If we can internalize such a force in the manner we have described, in terms of inertially augmenting the original system with more state variables and more parameters, then it is not too much an abuse of language to call the new variables we have introduced (and of course the parameters we need to couple them to the original system) *enzymes*. (This usage, however, embodies a confusion between active sites and the molecules that carry them [see *Life Itself*, section 11F].)

In formal terms, such augmented systems must be very heavily constrained, with all kinds of identical relations between the new variables and parameters we have added (i.e., the "enzymes") and the tangent vectors that govern change of state in the system. That is, the new variables are doing a "double duty": they define state in the larger system, and they also participate in operating on that state, in determining the rate at which such a state is changing.

Without going into details, these constraints are strong enough to be expressed in an abstract graphical language. Primitive examples of this are the familiar representations of intermediary metabolism, in which the arrows (representing enzymes) correspond to the inertial variables and parameters we have added to internalize impressed forces, and the vertices roughly correspond to state variables of the smaller open system on which the forcings are impressed.

The existence of such a graph expressing the constraints is in fact a corollary of internalizing forces impressed on open systems, not only in biology, but quite generally. To a large extent, the converse is also true, but that is not of immediate concern. Note that the graph looks very much like an aperiodic solid, and indeed it possesses many of the properties Schrödinger ascribed to that concept. The novel thing is that it is not a "real" solid. It is, rather, a pattern of causal organization; it is a prototype of a *relational model* (see *Life Itself,* chapter 5).

Since the larger system is itself open, the "enzymes" will themselves have sources and sinks. They are not present in the diagram, but without them, the enlarged system, represented by the graph, is generally *not stable* as a free system. If we want it to be stable, *we need more forces* impressed on the system to stabilize it. This is, roughly, where the Mendelian genes enter the picture.

In a nutshell, stabilization of this kind is attained by modulating the rates that the "enzymes" impose on the original open system with which we started. This, in fact, is precisely what the Mendelian genes do: they correspond to accelerations or decelerations of the rates at which "enzymes" themselves control rates. We may further think to internalize impressed forces of this kind in the same way we just internalized the "enzymes" themselves—namely, add still more inertial variables of state, and still more parameters to couple them to what we already have, to obtain an even bigger open system, and one that is even more heavily constrained than before. As before, these constraints are strong enough to be expressed in graphical language, but the kind of graph that arises at this level is much more complicated. Instead of two levels of "function" embodied in the distinction we have drawn between the arrows of the graph and its vertices, we now have three such levels (the original metabolites, the "enzymes" that force them, and now the Mendelian genes that force the "enzymes"). If the original graphical structures are indeed thought of as aperiodic solids, so too are the new ones, albeit of quite a novel type.

Unfortunately, even thus augmented, the resulting open systems are still not in general stable. We could repeat the process: posit new impressed forces to modulate the Mendelian genes we have just internalized and seek to internalize them by means of still more inertia (i.e., more state variables, more parameters to couple them to what is already in the system, and more constraints imposed upon them). At this point, we have a glimpse of an incipient infinite regress establishing itself.

The only alternative is to allow the sources and sinks for the internalized inertial forcers introduced at the Nth stage of such a process to have already arisen at earlier stages. A source for such an Nth-stage internalized forcer is *a mechanism for its replication,* expressed in terms of the preceding $N-1$ stages, and not requiring a new $N+1$ stage. Thus replication is not just a formal means of breaking off a devastating infinite regress, but it serves to stabilize the open system we arrived at in the Nth stage.

In biology, N seems to be a small number, $N = 2$ or 3, or perhaps 4 in multicellulars. But I can see no reason why this should be so in general.

Breaking off such an infinite regress does not come for free. For it to happen, the graphs to which we have drawn attention, and which arise in successively more complicated forms at each step of the process, must fold back on each other in unprecedented ways. In the process, we create (among other things) closed loops of efficient causation. Systems of this type cannot be simulated by finite-state machines (e.g., Turing machines); hence they themselves are not machines or mechanisms. In formal terms, they manifest impredicative loops. I call these systems *complex;* among other things, they possess no largest (simulable) model. The physics of such complex systems, described here in terms of the forcing of open systems (although they can be approached in many other ways) is, I assert, some of the new physics for which Schrödinger was looking.

When Is a Molecule a Mendelian Gene?

This was the real question Schrödinger was addressing in his essay, the inverse of the question Delbrück thought he answered by asserting that a gene is a molecule.

The question looks intriguing because, at its root, it embodies a Correspondence Principle between an "inertial" thing (e.g., a molecule) and a "gravitational" thing (a force imposed on an open system). But the question is much more context dependent than that; its answer involves not just inherent properties of a "molecule" in itself (e.g., "aperiodicity"), but also the properties of what system is being forced, and the preceding levels of forcing (of which "genome" is to be the last).

Thus very little remains of Schrödinger's simple cryptographic picture of "order from order," in which rigid molecular structures get transduced somehow into nonrigid phenotypic ones. Rather, the initial "order" appears as a pattern, or graph, of interpenetrating constraints, which determines what happens, and how fast it happens, and in what order, in an underlying open system. The arrows in such graphs, which I suggest constitute the real "aperiodic solid," are operators; they express "gravitational" effects on the underlying system. In terms of inertia, it is much more appropriate to speak of active sites than of molecules. The two are not the same.

Indeed, much simpler questions, such as, When is a molecule an enzyme? are hard to approach in purely inertial terms. These are all structure-function questions, and they are hard because a function requires an external context; a structure does not.

In a sense, if all we want to talk about is an active site (i.e., something gravitational), and we find ourselves talking about a whole molecule (i.e., something inertial), we run a severe risk of losing the site in the structure. There is much more inertia in a whole molecule than in a functional site. Unlike the impressed forces imposed from the environment of a system, which constitute an inertial defect, structure-function problems tend to involve a dual inertial excess of irrelevant information.

There is some new physics here too, I would wager.

What Is Life?

In this penultimate section, I shall review the Schrödinger question in the light of the preceding discussions, and in terms of a number of subsidiary questions either raised directly by Schrödinger himself or that have come up along the way.

Is What Is Life? a Fair Scientific Question?

My answer is, "Of course it is." Not only is it fair, it is ultimately what biology is about; it is the central question of biology. The question itself may be endlessly rephrased, but to try to eliminate it in the name of some preconceived ideal of mechanistic "objectivity" is a far more subjective thing to do than that ideal itself allows.

Does the Answer Involve New Physics?

Once we admit questions of the Schrödinger type, which treat an adjective or predicate as a thing in itself, we are already doing new physics. More formally, the old physics rests on a dualism between phases or states, and forces that change the states, which make the system "behave." Predicates, or adjectives, typically pertain to these behaviors, which are what we see directly. Moreover, the emphasis here is overwhelmingly skewed in the direction of what I have called the inertial aspects of a system (how it responds to forces) at the expense of its gravitational aspects (how it exerts forces). In biology, this shows up in terms of structure-function problems, where structure pertains to inertia, and function to gravitation.

Many biologists, indeed the same ones who would deny the legitimacy of the Schrödinger question, assert that function is itself an unscientific concept;[12] in effect, they assert there is only structure. Hence, biology can be scientific only insofar as it succeeds in expressing the former in terms of the latter. That is why Delbrück's argument, that a functionally defined Mendelian gene comprises a familiar chemical structure, a molecule, was received so enthusiastically, while the converse question (When can a "molecule" manifest such a function?), with which Schrödinger's essay is really concerned, was not even perceived.

Schrödinger's new physics, embodied generally in his initial question, and specifically in his appraisal of the relation between genes and molecules, rests in his turning our inertial biases upside down, or at least suggesting that inertial and gravitational aspects of material systems be granted equal weight. Once this is done, new physics appears of itself.

Is Biology "Marginal"?

Jacques Monod used this word in expressing his belief that organisms are nothing but specializations of what is already on the shelf provided by old physics, and that to claim otherwise was mere vitalism. He buttressed this assertion with his observations that organisms are in some sense rare and that most material systems are not organisms.

This kind of argument rests on a confusion about, or equivocation on, the term *rare,* and identifying it with *special* (see *Life Itself,* section 1A). An analogous argument could have been made in a humble area like arithmetic, at a time when most numbers of ordinary experience were rational numbers, the ratios of integers. Suddenly, a number such as π shows up, which is not rational. It is clearly rare, in the context of the rational numbers we think we know. But there is an enormous world of "new arithmetic" locked up in π, arising from the fact that it is much too general to be rational. This greater generality does not mean that there is anything vitalistic about π, or even anything unarithmetical about it; indeed, the only vitalistic aspects show up in the mistaken belief that "number" means "rational number."

Schrödinger's new physics makes an analogous case that organisms are more general than the nonorganisms comprehended in the old physics, and that their apparent rarity is only an artifact of sampling.

What Is This "New Physics"?

The new physics involves going from special to general, rather than the other way around. At the very least, it means going from closed systems to open ones, discarding specializing hypotheses such as closure conditions and conservation laws. There is still no real physics of such open systems, largely because the formalisms inherited from the old physics are still much too special to accommodate it.

Most significant, I feel, will be the shifting of attention from exclusively inertial (or structural) concepts to gravitational aspects. This can be expressed as a shift from concerns with material causations of behavior, manifested in state sets, to formal and efficient causations. As I have suggested, these are manifested in graphical structures, whose patterns can be divorced entirely from the state sets on which they act. The mathematical precedent here lies in geometry, in the relation between groups of transformations tied to an underlying space, and the abstract group that remains when that underlying space is forgotten. To a geometer, concerned precisely with a particular space, this discarding of the space seems monstrous, since it is his very object of study; but to an algebraist, it throws an entirely new perspective on geometry itself, since the same abstract group can be *represented* as a transformation group in many different ways (i.e., an underlying space restored, which can look very different from the original one from which the group was abstracted). In the same way, it would look monstrous to a biologist, say, to throw away his state spaces (his category of material causation, his "inertia") and retain an abstract graphical pattern of formal and efficient causation, but that is what is tacit in Schrödinger's concern with gravitation.

What Is Life?

The lines of thought initiated in Schrödinger's essay lead inexorably to the idea that life is concerned with the graphical patterns we have discussed. The formal metaphor I have suggested, namely, dissociating a group of transformations from a space on which it acts, shows explicitly a situation in which what is a predicate or an adjective from the standpoint of the space can itself be regarded as a thing (the abstract group) for which an underlying space provides predicates. This is analogous to the inversion of adjective and noun implicit in Schrödinger's question itself; as we saw at the outset, it involves partitioning an organism into a part that is its life and a part that is everything else. Seen from this perspective, the "life" appears as an invariant graphical pattern of formal and efficient causation, as a gravitational thing; the "everything else" appears in the form of material causation (e.g., state sets) on which such a graph can operate.

Such a system must be *complex*. In particular, it must have nonsimulable models; it cannot be described as software to a finite-state machine. Therefore, it itself is not such a machine. There is a great deal of new physics involved in this assertion as well.

To be sure, what I have been describing are necessary conditions, not sufficient ones, for a material system to be an organism. That is, they really pertain to what is not an organism, to what life is not. Sufficient conditions are harder; indeed, perhaps there are none. If so, biology itself is more comprehensive than we presently know.

What about "Artificial Life"?

The possibility of artificial or synthetic life is certainly left wide open in this discussion. However, the context it provides certainly excludes most, if not all, of what is presently offered under this rubric.

The first point to note is that, in open systems generally, analysis and synthesis are not inverse operations. Indeed, most analytic procedures do not even have inverses, even when it comes to simple systems or mechanisms. For instance, we cannot solve even an N-body problem by "reducing" it to a family of N_k-body problems, whatever k is. How much more is this true in the kinds of material systems I have called complex, a condition that I have argued is necessary for life? Indeed, no one has ever really studied the problem of when an analytic mode possesses an inverse, that is, when an analytic mode can be run backward, in any physical generality.

A second point is that what is currently called artificial life, or A-life, primarily consists of exercises in what used to be called biomimesis. This is an ancient activity, based on the idea that if a material system exhibits *enough* of the behaviors we see in organisms, it must *be* an organism. Exactly the same kind of inductive inference is seen in the "Turing Test" in artificial intelligence: a device exhibiting enough properties of intelligence *is* intelligent.

In this century, biomimesis has mainly been pursued in physical and chemical systems, mimicking phenomena such as motility, irritability, and tropisms in droplets of oils embedded in ionic baths. Previously, it was manifested in building clockworks and other mechanical automata. Today, the digital computer, rather than the analog devices previously employed, is the instrument of choice as a finite-state machine.

At root, these ideas are based on the supposition that some finite number of (i.e., "enough") simulable behaviors can be pasted together to obtain something alive. Thus that organisms are themselves simulable as material systems, and hence are not complex in our sense. This is a form of Church's Thesis, which imposes simulability as, in effect, a law of physics, and indeed, one much more stringent than any other. Such ideas already fail in arithmetic, where what can be executed by a finite-state machine (i.e., in an "artificial arithmetic"), or in any finite (or even countably infinite) collection of such machines, is still infinitely feeble compared to "real" arithmetic (i.e., Gödel's Theorem).

Schrödinger himself, in the last few pages of his essay, quite discounted the identification of "organism" with "machine." He did this essentially on the grounds that machines are rigid, essentially low-temperature objects, while phenotypes were not. This provocative assertion, more or less a small aside on Schrödinger's part, is well worth pursuing in the context of the material basis of artificial life.

Conclusions

Schrödinger's essay, published nearly a half-century ago, provides little comfort to an exclusively empirical view of biology, certainly not insofar as the basic question What is life? is concerned. On the contrary, it removes the question from the empirical arena entirely and in the process raises troubling questions, not only about biology but about the nature of the scientific enterprise itself. However, Schrödinger also proposed directions along which progress can be made. The consignment of his essay to the realm of archive is premature; indeed, it is again time that "everybody read Schrödinger."

NOTES

1. The paper that motivated Schrödinger was a joint work, which he himself referred to as "Delbrück, N. W. Timoféëff [Ressovsky] & K. G. Zimmer, *Nachr. a.d. Biologie d. Ges. d. Wiss. Göttingen,* vol. 1, p. 89. 1935." The main impetus of that paper was what in those days was called target theory (*treffertheorie*). It attempted to determine the cross section or physical size of "genes" by counting the number of mutations induced by standard doses of high-energy particles. It turned out (artifactually) that the answers obtained by such methods were of the order of molecular sizes. The stability arguments arose from attempts to understand or interpret these "target-theoretic" results.

2. This is the substance of Newton's Second Law of Motion. In this limited mechanical context, *behavior* means change of state (or phase). Force in this picture is always a function of phase alone; changing the force means changing that function, but the *arguments* of that function remain fixed, determined entirely by the system on which the force is acting. Thus in a strong sense, the same force acting on two different mechanical systems gets two different and unrelated descriptions. Even in mechanics, this view was strongly challenged, especially by Ernst Mach. The Machian challenge is rather closely related to the subject of this chapter.

3. The term *synergy* is used in different ways by different authors. It was used by Russian theorists, especially by I. Gelfand, S. V. Fomin, and their collaborators, to describe complicated coordinated neuromuscular activities like walking or running. It has been used, especially by H. Haken, in a way essentially synonymous with mechanical constraints. It connotes the control of many apparently independent degrees of freedom via a much smaller number of controls. Many different kinds of problems, such as the folding of proteins, can be regarded as synergetic in this sense.

4. At its essence, bifurcation theory concerns situations in which distinct systems that are close to each other according to one criterion may behave arbitrarily differently according to another. It thus concerns situations in which an approximation to a system, however good the approximation, fails to be a model or a surrogate for it. The original mathematical setting for bifurcation theory studies the interplay between a metric topology on a set and an equivalence relation ("similarity") on that set. Several of the essays in this volume deal with bifurcation-theoretic kinds of problems (see especially chapters 11 and 12). In biology, a wide variety of problems are of this type—for example, the distinction between micro- and macroevolution, and the extrapolation of experimental data from one species to another (Rosen 1959).

5. Physicist George Gamow was the first to suggest a cryptographic relation between linear copolymer sequences (primary structures) in DNA (or RNA) and polypeptides. Part of this was based on Schrödinger's use of the term *code* in his essay of 1944. Gamow's idea dated from early in 1953. It had apparently never occurred to any of the empiricists who considered themselves closest to the problem.

6. Intuitively, in physical terms, an open system is one that is exchanging matter and energy with its environment. Openness is harder to characterize mathematically; it is not the same as autonomy, the time-independence of imposed forces. It rather means the absence of symmetries or conservation conditions imposed on the system itself. These kinds of conditions tend to translate into the exactness of differential forms, which is very nongeneric (see chapters in part III).

7. Von Bertalanffy has become well known as the father of General System Theory. He came to develop this as an alternative to reductionist, Cartesian ideas, which he felt were not only scientifically inadequate for biology but had deplorable social and ethical side effects for humanity at large. His system-theoretic ideas were essentially relational in character. (See *Life Itself* for fuller discussions.)

8. Driesch, an eminent experimental embryologist, discovered in 1891 that if the two blastomeres arising from the division of a fertilized sea urchin egg are separated, each will develop into an entire organism and not into a half-organism. This behavior seemed to him so counter to any possibility of mechanical explanation, even in principle, that he invented a noncorporeal concept of *entelechy*, or "wholeness," to account for it. Behaviors such as this were often called equifinal, connoting that the same end-state is attained however much an intermediate state is perturbed or mutilated. (For an interesting discussion, see Waddington [1950].)

As far back as 1928, Bertalanffy identified "equifinality" with what we would today call the stability of point attractors in dynamical systems. He also recognized early that the equilibria of closed systems are not attractors at all. Hence, to a certain extent, Driesch was right, but he confused "physics" with "the physics of closed systems." Bertalanffy also recognized that open systems were not amenable to conventional, purely reductionistic modes of empirical analysis. This, ironically, led to the ever-increasing denunciation of Bertalanffy as a vitalist.

9. Turing's contribution is in *Philosophical Transactions of the Royal Society of London [B]* (1952;237:37). It should be noted that something more than equifinality is involved here: The stability of a steady state (point attractor) can change (in Turing's case, from a stable node to a saddle point) as a consequence of how open the system can be. This change of stability is manifested as a change from spatial homogeneity to spatial inhomogeneity; the latter behavior is the essence of self-organization. This is a bifurcation phenomenon between two classes of behaviors, each by itself equifinal.

10. Control Theory can be looked at, in a way, as the study of system behaviors as a function of state-independent, time-dependent forces. It thus involves a large conceptual departure from the completely state-dependent, time-independent kinds of forces envisaged by Newton (see note 2, above).

11. Roughly speaking, the transfer function of a system describes how it transduces input or forcing into output or behavior. It is the central concept of linear system theory. But it makes sense only for linear systems, and it does not generalize.

12. Expressions of the form "the function of the heart is to pump blood" or "the function of an enzyme is to catalyze a reaction" are regarded as teleological, hence vitalistic, hence unscientific (see *Life Itself,* section 5D). Recall the discussion of the Mendelian gene, initially defined purely in functional terms, and the concomitant denial of its reality on such grounds.

∾

CHAPTER 2

Biological Challenges to Contemporary Paradigms of Physics and Mimetics

The following remarks are intended to address two problems: (a) the role of contemporary physics in dealing with the nature and properties of living systems, and (b) the role of mimetic approaches (usually prefaced by the adjective *artificial*) in dealing with these same matters. Both approaches are offered as (quite distinct) ways of making biology scientific, or objective, by in effect making it something other than biology. And they are both, in a historical sense, ancient strategies; in their separate ways, they appear to embody a mechanistic approach to biological phenomena, whose only alternative seems to be a discredited, mystical, unscientific vitalism. They are alike in that they suppose biology to be a specialization of something inherently more general than biology itself, and the phenomena of life to be nothing but very special embodiments of more universal laws, which in themselves have nothing to do with life and are already independently known. In this view, whatever problems set biology apart from the rest of science arise precisely because organisms are so special.

One prevailing manifestation of such ideas is the naive *reductionism* that passes today as the prevailing philosophy underlying empirical approaches to organisms. The very word connotes that living things are special cases of something else, and that we learn everything there is to know about them by reducing them, treating them as mere corollaries of what is more general and more universal.

However, organisms, far from being a special case, an embodiment of more general principles or laws we believe we already know, are indications that these laws themselves are profoundly incomplete. The universe described by these laws is an extremely impoverished, nongeneric one, and one in which life cannot exist. In short, far from being a special case of these laws, and reducible to them, biology provides the most spectacular examples of their inadequacy. The alternative is not vitalism, but rather a more generic view of the scientific world itself, in which it is the mechanistic laws that are the special cases.

I shall document these assertions in a variety of ways, not least from a historical review of the repeated attempts to come to grips with the phenomena of life from the viewpoint that life is the special, and something else is the general. It is a history of repeated and ignominious failures. The conventional response to these failures is not that the strategy is faulty, but rather the presumption that organisms are so complicated (or to use the current jargon, complex) as material systems that the tactics of applying the strategy of reduction to them is simply technically difficult; it raises no issue of principle. Indeed, this very complexity is interpreted as a manifestation of how special organisms are.

On the other hand, if it is in fact the strategy that is at fault here, and not the tactics, the consequences are profoundly revolutionary.

On the Reduction of Biology to Physics

When I use the term *physics,* I mean contemporary physics—operationally, what is present in all the textbooks in all the libraries at this moment. This is different from what I have called *ideal physics,* the science of material nature in all its manifestations. Despite repeated historical attempts to replace the latter by the former, the two are not the same. Indeed, every advance in physics is a demonstration of something irreducible to yesterday's contemporary physics, and hence the nongenericity of it. Today's contemporary physics is no different, as biology clearly shows and indeed has always shown.

The statements by Einstein about "how primitive physics still is" and by Schrödinger about the need for "new physics" reflect the bafflement expressed two centuries earlier by Kant in his *Critique of Judgment:*

> For it is quite certain that in terms of *merely* mechanical principles of nature we cannot even adequately become familiar with, much less explain, organized beings and how they are internally possible. So certain is this that we may boldly state that it is absurd . . . to hope that perhaps some day another Newton might arise who would explain to us, in

terms of natural laws . . . how even a mere blade of grass is produced (the "Newton of the leaf").

Kant may perhaps be dismissed as a metaphysician, but he knew mechanics. Einstein and Schrödinger cannot be so dismissed; they knew their contemporary physics as well as anyone, and they knew what it could not do. To this day, today, the formidable powers of theoretical physics find nothing to say about the biosphere, nor does any physicist contemplating the mysteries of life speak of them qua physicist. This, I would argue, is because biology remains today, as it has always been, a repository of conceptual enigmas for contemporary physics, and not technical problems to be dealt with through mere ingenuity or the application of familiar algorithms. Somehow, the *life* gets irretrievably lost whenever this is attempted. Is this merely because we are doing it badly, or remain lacking in our data? This is hardly likely.

But if it is indeed an issue of principle that underlies the persistent separation between the phenomena of life and the material principles to which we strive to reduce them, the fault is not in the life, but in those principles. They are, in some sense, not generic enough to allow that reduction. The question for physics itself is then, Why not? The answers to this question would comprise the "new physics" that Schrödinger called for a half-century ago, a less "primitive physics" in Einstein's words.

Thus the complexity of organisms, in the conventional view, is interpreted as a measure of how special they are, considered merely as material systems. Complexity is measured in a system by counting the number of distinguishable ingredients it contains and the interactions that constrain them, by how complicated it looks to us. Thus in this view, it is generic for material systems not to be complex; most material systems are simple.

Moreover, it is conventional to suppose a kind of gradient of complexity in the material world, and to suppose that we move in this gradient from generic simplicity to sparse, nongeneric complexity by simple accretion, by a rote operation of adding one. Conversely, a reduction of one of these rare complex systems is merely an inverse rote operation of subtracting one, to get back the generic simple systems we presumably started with. Also presumably, these rote operations of adding and subtracting do not change the material basis of the systems themselves: simple systems are the same whether they are alone or whether they have been added into a larger one. This kind of context independence of simple systems is one central feature of scientific objectivity; its main corollary is that one must never pass to a larger system (i.e., a context) in trying to understand a given one but must only invoke simpler subsystems, specifically those that manifest a complete independence of context in their properties and behaviors.

These are extremely strong presumptions. They are worth stating again. First, simplicity is generic, complexity in any sense is rare. Second, simple systems, the generic ones, are entirely independent of any context. Third, the gradient from simplicity to complexity is only a matter of accretion of simple, context-independent parts, and the analysis of more complex systems is merely a matter of inverting the accretions that produced them.

Stated thus baldly, which they almost never are, these ideas do not look so self-evident. Each of them is seriously inadequate, and must be extensively qualified. Collectively, they serve to make biology unreachable from contemporary physics, which is based on them. At the same time, qualifying any of them would profoundly modify the conceptual basis of contemporary physics itself. This is why the phenomena of biology embody a foundations-crisis for contemporary physics, far more profound than any that physics has yet seen.

Indeed, it is more accurate to invert the presumptions. Namely, there is a sense in which complex systems are far more generic than simple, context-independent ones. Moreover, analysis and synthesis are not simple rote operations, nor are they in any sense inverses of one another. In short, the entire identification of *context-independence* with *objectivity* is itself far too special and cannot be retained in its present form as a foundation for physics itself.

Before proceeding further, we should try to characterize this physics. It comes in at least two varieties, which I call contemporary and ideal. The former keeps changing; we can hope that it approximates the latter better and better as time passes. The best characterization I have seen is Bergman's (1973) representation of Max Planck's view of the goal of physics, which appears at the beginning of chapter 5. I like it because it stresses, from the outset, a preoccupation with objectivity, the orthogonality of physics from any shred of the subjective. Indeed, this has been one of the main attractions of quantum theory since the time of Bohr—

the absence of any trace of what Einstein called *anschaulichkeit* (i.e., visualizability), or of what he regarded as "the free creation of the intellect," which alone can transform data into comprehension.

Biology, on the other hand, reeks of subjectivity by this criterion, in too many ways to count. Yet organisms are material systems; they project into the objective world just as any other material system does, and have been contemplated by physicists since ancient times with a mixture of horror and fascination. Their response has always been to try to suck the subjective life out of them; to reduce them to immaculately objective things designed to be orthogonal to them. We need only think of the large and curious literature on the famous "Maxwell Demon," which was not even alive, highly reminiscent of the response to Zeno's paradoxes.

My suggestion here is that this objective world, which constitutes the goal of physics, the ideal for which it strives, is in fact a highly nongeneric one, far too restrictive and specialized to accommodate things like organisms. Biology is not simply a special case in that world, a rare and overly complicated anomaly, a nongeneric excrescence in a generic world of objective things. To the contrary, it is that world itself that is nongeneric, and it is organisms which are too general to fit into it. This too counts as an objective fact, and it is one with which (contemporary) physics *must* come to terms, if it indeed seeks to comprehend all of material nature within its precincts. It cannot do this and at the same time maintain its claim to only allow objective things into it. Biology already will not pass through that extraneous filter, a filter which, ironically, is itself quite subjective in character.

A few remarks about mathematics are in order. In a certain sense, the mathematical universe is a subjective matter, residing entirely in the mind; it is precisely what Planck sought to exclude entirely from his objective world, both as subject and as object. On the other hand, this universe is also regarded as the very quintessence of objectivity: independent of the mathematician, independent of the world, independent of God, and beyond the scope of miracle in a way the material universe cannot be. Nevertheless, the physicist Wigner (1960) wrote that he marveled at the *unreasonable* effectiveness of mathematics in dealing with physical problems, the *unreasonable* way in which the mathematical realm and the material one seem to march in parallel. (In a way, it is not so surprising, since both realms comprise systems of inference or entailment, which in science is called causality.) And if this is true, then the following considerations are directly relevant to our present discussion.

For most of its history, mathematics regarded its own internal consistency as entirely self-evident, something that *had* to be so by the very nature of mathematics itself. But a series of rude shocks, mostly concerned with aspects of infinite sets and infinite operations, shook this faith and forced mathematics to turn back on itself, to try to entail such consistency of a mathematical system as a theorem, essentially as an effect or consequence of causes or conditions.

One way of trying to do this was, in effect, to preserve the familiar behaviors of the finite, in the infinite realms into which mathematics had long since moved. This crystallized, in the hands of people like Richard Brouwer and David Hilbert, into what has come to be called constructive mathematics. Its essence is to deny *existence* to anything that cannot be generated from a finite set of symbols by a finite sequence of rote operations (i.e., an algorithm). According to this view, most of the familiar properties and objects with which mathematics deals do not even exist; they are to be excluded from the mathematical universe. From this point of view, it is thus generic not to exist. Indeed, taking such a constructivist view, every infinite set requires a separate axiom to allow it to exist, its ticket of admission into the constructivist universe.

In effect, constructionism rewrote the concept of objectivity in mathematics; something was objective if, and only if, it was constructible. And as in physics, something constructible could be claimed to be independent of the "structuring mind" (using Bergman's words; see chapter 5). Nevertheless, it was hoped, and believed, that all of mathematics could somehow be crammed intact into this kind of constructible universe and thus recaptured pure, consistent, and free of paradox or internal contradiction.

These expectations were destroyed by Gödel's Incompleteness Theorem (1931). One way of looking at this theorem is as a demonstration of the nongenericity of constructibility, and hence of any notion of mathematical "objectivity" based on it. Stated otherwise, constructible systems are the rare ones.

Why are they so rare? Because there is not much, within the context of mathematics, that you can do in them; they are excessively impoverished in entailment. In retrospect, the idea behind them seems to be, If you cannot do much, you cannot get yourself into too much trouble.

This principle is perhaps most clearly visible in Bertrand Russell's discussion of these matters, which was couched in terms of what he called impredicativities, or vicious circles, which he argued were always associated with self-referential loops. He felt that assuring consistency was a matter of excluding all such impredicative loops, replacing impredicative things by equivalent but purely predicative ones. He was led to an ever-expanding "theory of types" in trying to do this, basically another approach to a constructible universe into which he tried to collapse all of mathematics. But as we have seen, it is simply not there; mathematical systems without impredicativities are too nongeneric, as mathematical systems in their own right, to accommodate them.

Likewise, the touchstone of objectivity in the Planckian sense is simply an assertion of the absence of impredicativities in material nature. It is presumed that any appearance of impredicativity in nature arises merely from the "structuring mind." But just as in mathematics, systems without impredicativities are excessively impoverished in entailment, in what can happen in them; that is precisely the consequence of how nongeneric they are.

I claim that, as material systems, organisms are full of such impredicativities. That is, they are far more generic than contemporary physics allows. Far from being special cases of that physics, they constitute vast generalizations of it. That is why physics has always had such trouble with them; it can look only at predicative pieces, which are not alive, and it cannot put those pieces back together to recapture the crucial loops.

Contemporary physics looks upon its objectivity, its constructibility or predicativity, with pride, as an expunging of any trace of teleology or finalism. But the impredicativities I have been discussing have nothing to do with teleology, any more than, say, Gödel's Theorem does. And just as a "constructive" mathematician, attempting to do Number Theory in his finitistic world, thereby excludes himself from almost all of "real" number theory (although he can keep indefinitely busy within that world), the contemporary physicist, through his presumptions of objectivity, likewise excludes himself from the outset from the basic phenomena of life.

It does not, from this perspective, help at all to try to replace an impredicativity by something sufficiently complicated within a predicative universe. Just as adding another axiom does not get around Gödel's Theorem, adding another particle will not close an impredicative loop.

The supposition that there is some constructive, predicative equivalent of such a loop, which can be reached by repeating a rote predicative operation sufficiently often, is a misconception.

The same considerations hold when it comes to "analyzing" a system such as an organism in reductionistic ways, ways that involve removing the individual components of an impredicative loop from the loop itself, breaking it open in a material sense. These are profoundly irreversible operations within a predicative context, and indeed, as far as the original system is concerned, they are an inexhaustible source of *artifacts*. Here, irreversibilities mean, among other things, that such reductions are not invertible; they are not the inverses of syntheses (at least, not in any predicative sense). That is, ultimately, why the origin-of-life problem is so hard. From this perspective, rummaging through a rubble of reductionistic fragments tells us little about either how the organism from which they came actually worked, or how it came to be; the "analysis" that produced those fragments cannot be inverted in either sense (again, not in any purely predicative context).

I earlier related the objectivity of contemporary physics to the presumed context-independence of its elements or units, and the forbidding of referring behaviors in a system to larger systems rather than to subsystems. This is just another way of talking about the exclusion of impredicativities, because it can be argued that a system is not known if its subsystems are not known, and knowing a subsystem requires knowledge of that larger system to which it belongs (an immediate impredicativity). The exclusion of such loops amounts to asserting that material systems can be approached only in a purely syntactic way; indeed, constructivity in mathematics was precisely an attempt to do mathematics by syntax alone. I would argue that contemporary physics is trying to grasp the material world by syntax alone. It did not work in mathematics; I believe that biology shows it will not work in the world of material nature either.

On Mimesis and Simulation

A major preoccupation of the search for objectivity in science is the removal of any residuum of a "structuring mind." One way to remove any possibility of such a residuum is to use something that, to the greatest extent possible, does not have a mind—to use a machine. Indeed, if a machine can "do" something, that is considered prima facie evidence both of the objectivity of what it does, and of how it is done. That is how the algorithm came to be the mainstay of constructive mathematics, and how the execution of algorithms, by means of syntactic engines, became tied to the notion of machine in the first place.

In fact, a common obsession with objectivity provides a basic conceptual link between ideas of reductionism and the apparently quite different strategy called *mimesis.* The latter is animated by an idea that things that behave *enough* alike *are* alike, interchangeable. In biology, such ideas go back to the earliest historical times and, indeed, are intimately related to the archaic concept of *sympathies,* as in "sympathetic magic." It is in fact quite incompatible with reductionism, and the modern fusion of the two approaches to biological reality, under the guise of objectivity, is one of the more ironic aspects of contemporary science.

In biology, mimetic ideas rest on the notion that, if we can produce a system manifesting enough properties of an organism, i.e., it behaves enough like an organism, then that system will *be* an organism. We shall come back to the idea that there is some threshold, embodied in the word *enough,* that serves to convert necessary conditions into a sufficient one, or that allows a system to be replaced by a sufficiently good simulacrum or mimic. Indeed, mimesis treats such individual behaviors much as a reductionist treats atoms, as syntactic units, into which the various protean activities of an organism can be reduced and then recombined by purely syntactic means. There is a supposition that if you assemble enough of these behaviors, you get all of them.

D'Arcy Thompson (1917), in his celebrated book *On Growth and Form,* gives some interesting glimpses into how mimesis was pursued in the last century. The emphasis then was mainly on trying to simulate cellular processes and behaviors, vaguely identified as motility, irritability, the ability to follow gradients (tropisms), and even a capacity to divide, in inanimate droplets. Today, the same spirit is manifested, not so much in these analog systems, but rather in digital ones, but the former still appear as scenarios in the literature concerned with origin-of-life questions.

We have seen that attempts to formalize mathematics produced only nongeneric simulacra of mathematics. Moreover, there is no threshold of "enough" to permit the replacement of the real thing by such a simulacrum, and no (syntactic, or predicative) way of pasting such simulacra together to recapture that real thing. Nevertheless, essentially by virtue of a misapplication of Occam's Razor, it is widely claimed that we can learn deep things about "real mathematics" by studying such algorithmic simulations; specifically, that if something can be done algorithmically, then that is how it was done.

In the physical world, the attempt to argue from a commonality of some behavior or property backward to a commonality of causal underpinnings, or more generally from an approximation to what it approximates, is widely known to be unsound. It underlies such things as the familiar Gibbs paradox; in mathematics, it is generally the province of (structural) stability. It involves the basic notion of *surrogacy,* and the possibility of moving observational or experimental data from an observed specimen to a larger class (where it constitutes what Eddington called *free information* about the unobserved ones [see chapter 13]).

The study of such simulacra, precisely because they are simulacra, has come to be prefaced by the adjective *artificial.* The most familiar is probably artificial intelligence, which is all that remains of a much more extensive and diffuse endeavor, variously called bionics or self-organization. The goal here was initially a joint attack on difficult problems, of an inherently interdisciplinary nature, arising from perception and cognition. But largely, this was swept away by an infatuation with the distinction between hardware and software, and the faith (analogous to that of constructibility in mathematics) that everything could be done with software alone. Indeed, for a long time, artificial intelligence degenerated into just the study of software, a concept extrapolated from machines on mimetic grounds, and one with no significance in the material world at all.

Today, that history is being repeated under the rubric of artificial life. Again, this finds its basis not so much in science but in mimesis, and the search for enough mimics to cross a threshold into life through software alone.

On Complexity

The idea that there is a threshold, predicatively crossable, between the animate and the inanimate, which in different ways underlies both reductionism and mimesis, was probably first forcibly articulated by John von Neumann. He called it complexity, and he argued that below this threshold, system dynamics could only decrease this complexity, but above it, complexity could actually increase. The earmarks of this increase are the characteristics of living systems: the ability to grow, develop, reproduce, evolve, learn, and so on. He never specified how this complexity was to be itself characterized or measured, but intuitively it had to do with cardinality and counting—with numbers of objective units, and of determinate interactions between them. Thus complexity was treated like a measurable, observable property of a system, and every system had a definite value of this system observable associated with it at an instant of time.

It is probably too late to change this terminology now, but I would much prefer to use a word like *complication* rather than *complexity*. Although von Neumann presented it as an objective system property, there is actually a very strong subjective component in it; in many ways, it is an artifact arising from a particular choice of coordinates. For instance, a continuing debate in ecology is the stability-diversity question, which roughly translates into the assertion that the more complex the ecosystem (i.e., the more species in it, and the more they interact), the more stable, or persistent, the system will be. In the most familiar models of population dynamics, *diversity* translates into how many nonzero terms there are in a certain matrix. But stability is a matter of the matrix invariants (i.e., the eigenvalues and eigenvectors of that matrix), and hence a matrix with lots of zeros (low diversity) can be far more stable than an apparently much more diverse (complex) one; diversity is an artifact of choice of coordinates, while stability is independent of coordinates. Likewise, a neural network can consist of many neurons and (axonal) interactions, but what it computes has a canonical Boolean form (like the Jordan canonical form for the matrices) in which that complexity disappears.

Systems that thus differ only by a subjective choice of coordinate system, but that give vastly different values of an observable measured on them, do not satisfy any kind of Planckian criterion of objectivity. Nor do they simply mimic each other. So what is this "threshold of complexity," manifested so dramatically in the behaviors of systems below it and above it?

A better clue to this term lies in the distinction we have already drawn between systems with irremovable impredicativities, and predicative simulacra of them. The issue here is not at all trying to decide whether two things are more or less complicated, within a predicative context, and to express this decision in terms of an observable defined in terms of counting or enumeration. The issue is rather the comparison of impredicative contexts with predicative ones, and with each other.

I would rather, then, call a system complex if it has inherent impredicative loops in it. Thus "real" number theory is in this sense complex; any constructible approximation to it, which cannot have any such loops, is simple. In other words, a system is thus complex if it has models which are themselves complex. A system is complex if it has noncomputable models. This characterization has nothing to do with more complication, or with counting of parts or interactions; such notions, being themselves predicative, are beside the point.

It should be noted that there are many deep parallels between the dichotomy we have drawn between simple (predicative) and complex (impredicative), and the dichotomy between the finite and the infinite. Just as "infinite" is not just "big finite," impredicativities are not just big (complicated) predicativities. In both cases, there is no threshold to cross, in terms of how many repetitions of a rote operation such as "add one" are required to carry one from one realm to the other, nor yet back again.

The conventions on which contemporary physics rest amount to asserting that the world of material nature, the world of causal entailment, is a predicative world. It is a world of context-independent elements, with a few finitary operations, in which impredicativities cannot arise. This is what makes it objective. But this objectivity is bought very dearly: its cost is a profound nongenericity of that world, an impoverishment of what can be entailed in it. Most profoundly, it is a world in which life does not exist. But life does exist. The world of material nature is thus not in fact a predicative world. That is the Foundation Crisis faced by contemporary physics—that the world to which it aspires is a complex one, not a simple one.

CHAPTER 3

ᜤ

What Is Biology?

In this chapter, I had intended to consider only the legitimacy of identifying two very differently defined things: a "Mendelian gene," defined in indirect functional terms via its manifestations in larger systems (phenotypes), and an intrinsic structural feature (sequence or primary structure) of a polymeric molecule. As it turned out, this question could not be readily separated from its own, deeper context, one that goes to the very heart of reductionism itself. I have touched already on many of these issues in *Life Itself* (1991), in dealing with the theoretical basis of organism, when I argued that reductionism provides a flawed and inadequate strategy for biology, that its modes of analysis have little to do with how organisms are actually put together, or with how they work. In particular, they cannot be "run backward," because they rely on purely syntactic, algorithmic, machinelike modes of entailment that are in themselves extremely limited, nongeneric.

One corollary of this approach is that organisms, as material systems, raise profound questions at the foundations of physics, the science of material systems, itself. Simple-minded reductionism (see, e.g., Monod 1971) claims it is a form of vitalism to assert the possibility of learning anything essentially new about physics from a study of organisms. There are many reasons, not a few drawn from the historical development of physics itself up to the present, for not taking such condemnations seriously. In a cognate area, for example, it is a theorem that mechanistic, algorithmic, syntactic approaches to mathematics are simply too feeble to capture any but a vanishingly small fragment of "real" mathematics, but there is nothing vitalistic about asserting that algorithms are the special cases in mathematics, rather than the reverse.

All this is well exemplified in the sequence hypothesis, which is today widely regarded as completely embodying the mechanization of organism, and as an illustration of the mechanical character of physics itself. This last, of course, is something that no physicist today would claim, at least without enormous qualification. So the sequence hypothesis, apart from its direct claims, is itself directly tied to some of the deepest foundational issues in science. Most interesting, these issues are not themselves empirical issues, to be decided by experiment or evidence, any more than the epicycles of Ptolemy could be refuted thereby, or than Gödel's Incompleteness Theorem could be established. Yet the issues involved are utterly basic.

The Mendelian Reduction of Phenotype

Mendel's initial concerns were entirely tied up with phenotype (from a Greek verb meaning to show, hence that which is visible, tangible, directly observable) or soma. He worked with pea plants, which manifested visible somatic "characters": tall or short, wrinkled or smooth cotyledons, yellow or green cotyledons, etc. In a sense, each of these characters constituted a modifying adjective for the pea plant itself. But the whole plant is the noun, and adjectival characters like *tall* or *short* mean little in themselves—they acquire meaning only relative to the noun they modify.

It was Mendel's idea that each such discrete phenotypic character required an associated hereditary factor, which in some sense controlled or forced it, as described in chapter 1. Lying rather deeper is the tacit idea that the pea plant itself is only an aggregate or sum of its individual phenotypic characters. That is, the plant is nothing but a bundle of its discrete adjectives, which all together are *equivalent* to the noun they modify and, indeed, more than equivalent to it, actually *identical* with it. The equivalent assertion in mathematics would be that the totality of necessary conditions for something comprise a sufficient condition.

Thus underlying Mendel's approach is the idea that a pea plant, and by extension any organism, can be fractionated into a family of discrete somatic characters, each of which can be accounted for by hereditary factors of its own. Doing this, feature by feature, character by character, would thus account for everything about the whole organism, including its very life.

This approach constitutes a very strong form of reductionism, but it is a very different kind of reductionism from that asserted today. For instance, it argues backward from characters to factors, because the fractionation on which it is based arises at the phenotypic end. Today we prefer to argue forward, from factors to characters, because today's ideas of fractionation arise at the other end. Nevertheless, it is simply presumed that the two are essentially equivalent, in the sense that there exists a syntactic dictionary (or code) that algorithmically translates the Mendelian fractionation into the other one.

Let us then briefly turn to a consideration of this other reductionistic fractionation of an organism (e.g., a pea plant), which is to be compared to the Mendelian one.

Molecular Biology

It is a direct corollary of the Atomic Theory that, as a material system, an organism could be viewed in quite a different way from the Mendelian one just described. Namely, an organism was to be understood in terms of the population of particles of which it is composed. Hence, its properties as a material system were to be referred back to those of its constituent particles, the factors and characters together. An early version of this form of reductionism was the Cell Theory, where "particle" was identified with "cell." Today, *particle* generally means *molecule.*

This is, of course, the idea underlying the molecular biology of today. An organism is now to be fractionated, not into its somatic characters and their underlying hereditary factors, but into a bundle of molecular fractions, in terms of which any other fractionation, such as the Mendelian one, has to be subsumed.

The analysis of an organism (or indeed, of any other material system) into a bundle of discrete particulate fractions thus expresses that bundle as a surrogate organism. This surrogacy is based, not at all on any commonality of actual behavior between the organism and its fractions, as the Mendelian one was, but entirely on the fact that both of them consist of the same particles.

Since particle-based fractionations, or analyses, are applicable to any kind of material system, they are regarded as more fundamental than any other. Hence, any other mode of analysis, such as the Mendelian, must be merely a special case of them. It is merely a question of expressing the analytic units of the latter in terms of those of the former. In our case, it is a question of expressing the characters and factors of a Mendelian analysis directly in terms of molecular fractions; this in itself constitutes the dictionary, or code, between the two analyses. More particularly, it must subsume any idea of a coding relation between the Mendelian characters themselves and the hereditary factors that control them—a quite different kind of coding.

Today, the problem is considered solved by the identification of everything genetic in an organism with a fraction of molecules called DNAs; all other fractions are epigenetic (I believe it was David Nanney who first employed that term in this fashion). Hence the Mendelian phenotypic characters are in there. It is just a question of finding them and identifying them. Moreover, whatever they are, it is supposed that the relation between the genetic fraction and its complement, the epigenetic fraction, is essentially the same as between the program to a machine and the execution of that program as a sequential diachronic process extended in time. This is in fact the substance of the sequence hypothesis.

Factorizations

The two modes of reduction I have been discussing (namely, the Mendelian and the molecular) give us, on the face of it, two distinct ways to visualize an organism (e.g., a pea plant). Here, I begin to explore the assertion that these modes are not in fact different but merely express the same things in different languages—hence that there is a dictionary, or code, for translating between them. Indeed, that this dictionary is a purely syntactic thing, a function only of the languages, and quite independent of what the languages actually refer to.

According to my more extensive discussion in *Life Itself,* each of these languages constitutes a kind of mapping, an *encoding,* from a material system Ω (say an organism, such as a pea plant) into a formal system, a symbolic system or language in which one can draw conclusions from premises, in the broadest sense a mathematical system. Such encodings must satisfy a strong condition, namely, that inferences drawn from premises encoded from Ω will decode back into what is

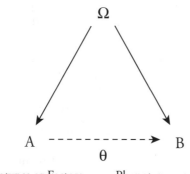

$$\Omega$$

A - - - - - - - - → B
θ

Genotypes or Factors   Phenotypes or Characters
(Microstates)   (Macrostates)

FIGURE 3.1

actually happening in Ω. More formally, that there is a correspondence, or homomorphism, between the causal structure in Ω and the inferential structure in our language, preserved by the encoding. This is the essence of what I called a *modeling relation.* There is of course much more to it, but the idea that there is a homomorphism between causal processes in the system Ω we are talking about, and the inferential structure of the language we are using to talk about it via our encoding, will suffice.

In the present situation, we have two apparently distinct ways of encoding the same system Ω: a Mendelian way based on phenotypic fractionation, and another way based on molecular fractionation. This is expressed in figure 3.1, a diagram very familiar in mathematics.

The basic question in any such diagram is whether we can draw another arrow of the form $\theta : A \to B$ (i.e., between our two different encodings of Ω) that will make this diagram commute. If so, then such an arrow constitutes our dictionary between the encodings; it is like a modeling relation between the encodings, and it expresses a homomorphism between the inferential structures of the languages, independent of Ω.

There are several ways of talking about such a situation. When such a θ exists, we may say that we have *factored* the encoding $f : \Omega \to B$ through A, in the sense of being able to write $f = \theta g$. We may also say that we have *lifted* the mapping f to the mapping g. In either case, we can put an equivalence relation R on A, essentially defined in terms of what the map θ can or cannot discriminate in A, so that the set B is just A reduced *modulo* this equivalence relation—formally, B = A − R. Finally, we can call the elements of B *macrostates,* and those of A the corresponding *microstates.*

It is extremely unusual, nongeneric, for factorizations of this kind to exist. When they do exist, we can say we have reduced one of the encodings to the other. Otherwise, the encodings are mutually irreducible; they are of different generality; they are complementary, in the sense of Bohr. When this happens, neither encoding of Ω suffices to fully capture, in its modes of entailment, all of the causal processes that govern what happens in Ω itself. I shall consider this kind of situation again in the final section of this chapter.

With this general backdrop, we return to our specific situation— namely, if Ω is an organism (e.g., a pea plant), describable by two apparently different encodings (e.g., the Mendelian one and the molecular one), can we factor the former through the latter, or can we not? If we can, how can we do it? If we cannot, then neither description is complete; there are things going on in Ω that neither encoding captures. The answers to such questions are profound enough for biology, but they would be even more so for physics, because it is one of the basic tenets that a sufficiently extensive particulate fractionation comprises a "largest model," through which every other encoding effectively factors.

Searching for θ: The Boundary Conditions

Thus it is generally supposed that any relation between the molecular fractionation and the Mendelian one must proceed from a fixed identification between the Mendelian hereditary factors and the DNA fraction. This is indeed the basis of any assertion of the complete reduction of biology to physics. It is further believed that, once this identification is made, the rest is an entirely technical matter, involving no further questions of principle.

But, as I have noted, the Mendelian gene and the molecular-biological one are very differently defined and measured. The former is a forcer of (epigenetic) dynamics, visible entirely in its phenotypic effects; since the days of Mendel, such factors are observed by comparing a standard phenotype (the wild type) with another, differing from it in one allele. The latter, on the other hand, is an intrinsic structural feature of a molecule, entirely independent of any larger system; this is precisely what is supposed to make it objective.

In physical terms, such an identification means equating a particle with a force that moves particles. At the very beginning of Newtonian particle mechanics, the sharpest possible distinction was drawn between phases of a particle, which change in time due to forces impressed on them, and the forces themselves. This duality is manifested in two entirely different kinds of laws: the Newtonian Laws of Motion expressed how a material particle changes phase in terms of the impressed forces that push it, while the Law of Gravitation expressed how a particle can exert forces on other particles.

How a particle moves when pushed by a force is, roughly speaking, the province of its inertia, embodied in mechanics in a specific parameter called inertial mass. How a particle pushes another is embodied in another, quite differently defined parameter, called gravitational mass. Experiment showed that these two different parameters turn out to be numerically equal—inertial mass = gravitational mass. The peculiar nature of this identification of two very different things was ultimately perceived by Einstein as one of the deepest things in physics; in his hands, it became a Principle of Equivalence between inertia and gravitation, and it eventually provided an essential cornerstone of General Relativity. The equivalence between inertia and gravitation thus paved the way for a profound extension of mechanics itself, something that could not be encoded into classical mechanics at all.

In the present case, intrinsic properties of a DNA molecule, such as its sequence, are inertial properties. The forces it must exert on other, larger systems around it, on the other hand, are gravitational in nature. Just as in mechanics, identification of the two amounts to an assertion of a Principle of Equivalence, which directly transduces inertial properties into gravitational ones.

A limited attempt to do this kind of thing in a chemical context, although never expressed in these terms, was developed a long time ago under the title Absolute Reaction Rate Theory (Glasstone, Laidler, Eyring 1941). The idea here is to express gravitational things, such as kinetic rate constants in chemical reaction systems, in terms of inertial aspects of the reactants. Actually, its provenance is far wider. As these authors remark, "The theory of absolute reaction rates is not merely a theory of the kinetics of chemical reactions; it is one which can, in principle, be applied to any process involving a rearrangement of matter; that is to say, to any 'rate process.'"

At issue here is the conversion of synchronic, inertial properties into diachronic, dynamical, epigenetic ones. But dynamics raises a host of deep questions of its own. These were manifested early, in terms of Henri Poincaré's attempts to deal with classical mechanical N-body problems as rate processes. He established a program of trying to determine basins of attraction, and the attractors themselves, from the rate constants of other parameters in the dynamical laws, without actually solving these equations. This can be accomplished for $N = 1$ or 2, but not generically beyond that. Moreover, such N-body problems cannot be approached recursively; that is, an N-body problem cannot be solved, in any sense, by solving (N_k)-body problems. This is already a classical failure of reductionistic fractionation—that you cannot approach a three-body problem by separating a three-body system into, say, a two-body system and a one-body system, and solving those. In a sense, there is not enough "gravitation" in the fractionated system to allow this, although all the original "inertia" is still there.

In the present case, a simple identification of DNA sequence and Mendelian gene means a direct confrontation with all these issues, at the very least. Specifically, we require the following:

1. a principle of equivalence, to effectively transduce inertial sequences into gravitational, epigenetic forces;

2. an absolute rate theory to turn these into explicit epigenetic equations of motion (i.e., to attach tangent vectors to states or phases in larger systems);

3. a way to determine global basins of attraction, and the attractors themselves, directly from the gravitational parameters ("rate constants"), for these must be the Mendelian phenotypes.

This is the very least that is required to establish a factorization of the Mendelian fractionation through the molecular one, subject to the boundary condition that inertial DNA sequence is to be identified with the gravitational Mendelian hereditary factor.

Some General Comments

As previously noted, the general philosophy of material reductionism of a physical system argues that any model of it must factor through a "biggest model," supplied by dissecting or fractionating the system into ultimate constituent particles. The claim is, then, that such a biggest model at least always exists. Thus such a biggest model is like a free structure in algebra, of which any other model is a quotient structure. Accordingly, the entire scientific enterprise consists of lifting a particular model to this biggest one. It is also believed, although this is an independent matter, that such a lifting is of a purely syntactic, algorithmic, effective character, independent of what is being modeled; it can in principle be described to, and executed by, a machine via a sufficiently elaborate program.

The focus here has been on one manifestation of these ideas, on trying to factor the Mendelian reduction through the molecular one. The latter is presumed to be the largest model, and hence it is argued that a factorization must exist. Moreover, it is believed that this presumptive factorization will necessarily identify DNA primary structure (sequence) with Mendelian hereditary factors.

However, actually producing such a factorization is most difficult, even in the inorganic realm. We can, in fact, exemplify these difficulties in the notorious protein folding problem, where there is presumably no biology at all. Here, what DNA sequence codes for is only a primary structure, basically a denatured polypeptide, devoid of any tertiary structure. To assert that the primary structure alone is sufficient to dynamically generate the tertiary constraints (i.e., to assert the infinite elasticity of the folded structure) is already to assert a transduction of synchronic, inertial information into the rates governing the diachronics of folding. We must then convert the asymptotic features of the folded structure, which take the form of "sites" (the phenotypes of the folded protein) again into rates in external epigenetic dynamics, in ways quite unpredictable from primary structures alone. Moreover, if we try to "decode" these sites, in themselves, back to their DNA sequences, we obtain things that have no primary structure at all.

Such considerations, and many others not mentioned here, raise the following possibilities, each successively more radical:

1. There exists a factorization Θ of Mendelian through molecular, but it does not satisfy the boundary condition DNA sequence = Mendelian gene.

2. There exists no factorization at all, but there does exist a largest model through which both (differently) factor. That is, the Mendelian and the molecular pictures are complementary, in the sense of Bohr.

3. There is no largest model.

The first possibility says that the Mendelian picture cannot be "lifted" to the molecular one via an immediate, exclusive identification of DNA sequence and Mendelian factor, but it can nevertheless be lifted. The second, more radical possibility says that the Mendelian picture cannot be lifted at all, at least not within the confines of contemporary views about reductionistic particles and their interactions. The final, most radical possibility is closely related to what I have called complexity. It says basically that organisms cannot be completely formalized; that they have nonalgorithmic, noncomputable models, and hence no largest, purely syntactic one.

Put briefly, all of these possibilities assert, in successively more radical ways, that all fractionations constitute *abstractions;* they lose essential information about the original intact system. They simply differ in what information they lose. This holds true for the molecular reductions of physics as well, the presumptive largest models, and hence not regarded as abstractions at all; that indeed is the meaning of the assertion that every other model or fractionation must factor through the biggest one.

All abstractions proceed by "hiding variables." But the variables so hidden in a particular abstraction show up when the behavior of the abstraction is appropriately compared to the corresponding behavior of the original system. They tend to be swept under the rug, into an all-encompassing term, *organization.* This organization can itself be characterized further, in biological contexts, and such characterizations involve what I have called relational ideas.

Schrödinger, in *What Is Life?* (1944), was emphatic that biology would require "new physics" to accommodate it. This assertion is now called vitalistic. However, attempts to factor Mendelian fractionations through molecular ones (i.e., to factor the provenance of biology through the provenance of physics), as described in this chapter, vividly indicate how right he was.

Part II

❧

ON BIOLOGY AND THE MIND

THE MIND-BRAIN problem is somewhat apart from my direct line of inquiry, but it is an important collateral illustration of the circle of ideas I have developed to deal with the life-organism problem. I do not deny the importance of the mind-brain problem; it was simply less interesting to me personally, if for no other reason than that one has to be alive before one is sentient. Life comes before mind, and anything I could say about mind and brain would be a corollary of what I had to say about life. That is indeed the way it has turned out.

It is, in a sense, fair to say that the brain is widely, if tacitly, regarded almost as if it were an organism in its own right, or a suborganism of the larger one to which it belongs. It is built of cells (neurons) and tissues, just as an independent organism is, and functions almost as a symbiont in its relation to the organism as a whole. However, it also thinks. So the question arises, to paraphrase Schrödinger again, What is thought? And the answer to this question is sought in the anatomy and the physiology of the brain, of this semiautonomous "neural organism." Hence, the endeavor to reduce mind to brain.

So the problem revolves around reductionism again, that same reductionism that seeks to express life itself in terms of a dance of isolable chemical fractions. In the following chapters, I pose this reductionism in terms of the concept of objectivity, because this is reductionism's mainstay. It arises in its sharpest form in the mind-brain problem, because thought is subjective, personal, unscientific, as von Neumann says, "by its very nature." By virtue of its inherent attempt to objectify the subjective, the mind-brain problem turns out to have an intrinsic bite, which even the life-organism problem does not manifest.

So, at root, the chapters in this part are concerned with objectivity.

As a whole, they argue that attempts to identify objectivity with algorithms and their execution have been a profound mistake. Chapter 4 deals with mathematics: on one side, the impregnable bastion of objectivity, but on another, an entirely subjective creation of mind. I briefly review some 2,500 years of mathematical history, culminating in the most recent attempts to reduce mathematics to algorithms: that is, to express all that is objective in mathematics as a special case of what is algorithmic or constructible. Alas, it is the other way around: all experience shows that it is the algorithms that are the (excessively) special cases.

In chapter 5, I analogize the presumed separability of objective and subjective in the material world with the possibility of similarly separating constructible or algorithmic mathematics from the rest of it. Chapter 4 does not go this far but simply describes the history of one idea (commensurability), an idea redolent of constructibility and objectivity, and the successive failures arising from attempts to prop it up. I draw the reader's attention to one conclusion of this investigation: that most of the troubles I describe arise from trying to pull a power set back into the original set, as reductionism (and objectivity) essentially requires.

The use of mathematics to illustrate a point about material nature arises quite naturally here. In fact, mathematics has played a number of key roles in the history of the mind-brain problem. Historically, the crucial idea, which in one way or another has imbued all modern thought about the problem, was the conception of the neural network dating from the work of Nicolas Rashevsky in the early 1930s. He converted a relatively routine investigation of peripheral nerve excitation into an approach to the central nerve, the brain itself, by arraying excitable elements into networks and observing that such networks could behave in "brainlike" ways; for example, they could think logically. A decade later, his approach was recast in algebraic, Boolean terms by McCulloch and Pitts (1943). In this form, the ideas meshed well with independent work by Turing on algorithms and machines (1950), motivated by the kinds of mathematical foundational problems that Gödel and many others were addressing. They also meshed well with contemporary technological developments, such as switching circuits, information theory, and, ultimately, cybernetics. Some of this spilled over into mathematics itself, in the form of, for example, recursive function theory and automata theory. Meanwhile, over the years, many other kinds of biological processes, including operon networks (and biochemical control processes generally) and the immune system, were subjected to the same kind of treatment. All told, the neural network has proved one of the most fertile and durable ideas in biology, and I emphasize that it was Rashevsky's idea.

At any rate, the mathematical correlates of neural nets indicated a tantalizing relation between brain and one of the most sublime products of thought, mathematics. It indicated a way to express the objectivity of mathematics and reconcile this with the inherent subjectivity of thought itself. So, in this context, it is not so far-fetched to regard developments within mathematics as closely cognate to, and bearing directly on, such material questions as the mind-brain problem. At root, this treatment merely exemplifies what, in *Life Itself*, I called Natural Law. We shall return to these ideas subsequently.

Chapter 5 contains my first explicit attempt to talk about the mind-brain problem. It was commissioned by a group of colleagues at the Karolinska Institute in Stockholm, who had organized a seminar devoted to that problem. Realizing early that objectivity was the central issue, and that the objectivizing of mind was the dual, reverse aspect of theory (i.e., the subjectivizing of nature, to the point where the mind can grasp it), I was struck by the obvious impredicative loop that arises from putting these two things together. In the quotation that begins the chapter, Planck asserts that it is nothing less than "the task of physics" to draw a sharp line between the world and the "structuring mind," and to stay entirely on its own side of this line. I recast this as the separation between nature and an observer of nature; objectivity then becomes transmuted into the presumption that the observation process itself plays no causal role in the result of the observation (i.e., it answers no question Why? about it). That is, objective means a causal independence from any such larger context. As such, it embodies the reductionistic idea that one must never invoke such a larger context as an explanatory principle; one must never refer to a larger system, but only and ever to subsystems.

Of course, this presumed independence of observed from observer fails already in quantum mechanics and as such constitutes the heart of the "measurement problem" therein. I conclude that there is no such line between subject and object: to suppose such is tantamount to postulating a simple world. And, as I argued in *Life Itself*, whatever our world may be, it is not simple. Hence, by extension, a reductionistic excursion into the mind-brain problem in the name of objectivity merely exposes the inherent limitations of reductionism itself.

Chapter 6 was intended as a sequel to chapter 5 and was prepared under the same auspices. Here, I introduce a somewhat different angle from which to view the same overall problem, expressed in the deliberately provocative title, "Mind as Phenotype." It reflects my long-time preoccupation with the biological dualism between phenotypes and underlying genotypes. My approach to these problems dates from my invocation of Aristotelian causal categories in discussing this dualism, and the identification of genome (in the sense of species determining) in terms of formal causation.

Nobody can say where biological phenotype begins or ends. For many, it ends with protein synthesis. For others, it ends with five fingers, dappled coats, and notched leaves. For still others, it ends with spider webs and termite mounds, with mating rituals, courtship displays, language, and behavior. Accordingly, no one knows what genome is: it depends entirely on the context of phenotype which is presumed (another nice impredicative loop). Is mind, then, a legitimate part of an individual's phenotype? If not, why not? If so, what is the genotype underlying it? The first issue is the indeterminate nature of these basic terms of contemporary biology, about which there is so much argument and so little understanding.

I then describe two different, popular (and equally unsuccessful) strategies for exploring such questions. The first I term *mimesis*, an attempt to argue from a homology between behaviors (phenotypes) to a homology of causal underpinning (especially of genotypes). The other is reductionistic fractionation, the replacing of a highly distributed set of behaviors by looking at a set of parallel fractions (i.e., by something with a vastly different phenotype) but nevertheless asserting a commonality of genome. Although these approaches proceed from entirely different (indeed, incompatible) presumptions, they are alike in their invocation of objectivity and their reliance on algorithms.

Chapter 7 involves a more detailed discussion of mimetic strategies, using the Turing Test as a case in point. I believe it stands by itself, but the reader is advised to be aware of the larger contexts provided by the other chapters in this section.

Chapter 8 is probably the most recent essay in this volume. I restate the issues raised earlier but now introduce a presumed hardware-software dualism, an extrapolation drawn primarily from our own technologies, and I superimpose it upon a more detailed analysis of reductionistic fractionation.

The thrust of these chapters, and, indeed, of most of this volume, is primarily critical. I do not explicitly offer a solution of my own to the mind-brain problem: I only indicate why I do not look for a solution from either reductionistic or mimetic directions. Nevertheless, I regard the mind-brain problem, and the other cognate problems I discuss, as touching on some of the deepest issues of science and beyond, issues that cannot be addressed without making some deep reappraisals of what science is and what it is about. Moreover, I feel I have indicated sufficiently, in *Life Itself*, the positive directions along which more tangible solutions to such problems are likely to lie.

CHAPTER 4

ᐛ

The Church-Pythagoras Thesis

It is my contention that mathematics took a disastrous wrong turn some time in the sixth century B.C. This wrong turn can be expressed as an ongoing attempt, since then, to identify effectiveness with computability. That identification is nowadays associated with the name of Alonzo Church and is embodied in Church's Thesis. But Church was only among the latest in a long line going back to the original culprit. And that was no less than Pythagoras himself.

From that original mistake, and the attempts to maintain it, have grown a succession of foundation crises in mathematics. The paradoxes of Zeno (described later in this chapter), which it has been said retarded the development of the calculus by a thousand years, were an immediate outgrowth of the thesis. The plague of divergent sequences, which nearly wrecked mathematics in the eighteenth century, constituted the next outbreak. We are presently mired in yet another one, ironically the one that Church was provoked to help resolve by rearticulating that original mistake in modern dress.

The impact of that wrong turn, made so long ago, has spread far beyond mathematics. It has entangled itself into our most basic notions of what science is. It might seem a far cry from the ultimately empirical concerns of contemporary science to the remote inner world of mathematics, but at root it is not; they both, in their different ways, rest on processes of measuring things, and on searching for relations ("laws") between what it is that they measure. From this common concern with measurement, concepts pertaining to mathematics have seeped into epistemology, becoming so basic a part of the adjective *scientific* that most people are quite unaware they are even there.

From earliest times, mathematics has had an invincible aura of objectivity. Mathematical truth was widely regarded as the best truth—independent of the mathematician, independent of the external world, unchangeable even by God himself, beyond the scope of miracle in a way that the material world never was. Science has always craved and coveted that kind of objectivity. That is what the thesis seems to offer. But it has been a dangerous gift.

Pythagoras

The name of Pythagoras is commonly associated not only with the history of mathematics but with its involvement in a form of occult numerical mysticism, which purely mathematical considerations were employed to justify. There is certainly a cosmogonic aspect to Pythagorean doctrine, in addition to a religious aspect. Over against these received notions, consider the following, from W. K. C. Guthrie (1962):

> [Pythagoras'] father Mnesarchos of Samos . . . is described as a gem-engraver, and it would be in accordance with regular Greek custom for Pythagoras to be trained in his father's craft.   (p. 173)

> Aristoxenus, the friend of fourth-century Pythagoreans, wrote in his treatise on arithmetic that Pythagoras derived his enthusiasm for the study of number from its *practical applications in commerce*. This is by no means an improbable supposition. The impact of monetary economy, as a relatively recent phenomenon, on a thoughtful citizen of mercantile Samos might well have been to implant the idea that *the one constant factor by which things were related was the quantitative*. A fixed numerical *value* in drachmas or minas may "represent" things as widely different in quality as a pair of oxen, a cargo of wheat, and a gold drinking-cup.   (p. 221; *emphases added*)

Even more surprising, perhaps, is the following:

> Pythagoras may have both introduced and designed the unique incuse coinage which was the earliest money of Croton and the neighboring South Italian cities under her influence. This is a coinage which excites the enthusiasm of numismatists by its combination of a remarkable and difficult technique with outstanding beauty of design, and Seltman

claims its sudden appearance with no evolutionary process behind it postulates a genius of the order of Leonardo da Vinci. . . . As the son of an engraver, he would himself have been a practising artist, and of his genius there can be no doubt. One begins to appreciate the dictum of Empedocles that he was "skilled in all manner of cunning works."

(p. 176)

These intriguing possibilities cast an entirely new light on the personality behind the adjective *Pythagorean.* It is also curious to note that, some two millennia later, Isaac Newton was to concern himself with coinage as well.

Thus there was a motivating concern for *quantitation* and the treatment of numbers as *values* (i.e., as a way of dealing with qualities). It is only a short conceptual step from this concern to the more modern concept of *measurement,* and to the idea of a (numerical) value as something inherent in what is being measured, not merely a convenient convention.

The Pythagorean concept of harmony was an essential step in developing relations between numbers and qualities:

The word *harmonia* . . . meant primarily the joining or fitting of things together, even the material peg with which they were joined . . ., then especially the stringing of an instrument with strings of different tautness (perhaps thought of as a method of *joining* the arms of a lyre . . .), and so a musical scale.
(Guthrie 1962:220)

In such a context, there is an essential connotation of design, or craft, in the word *harmony,* a connotation much stronger than the word has in its present usage.

To Pythagoras has always gone the credit for discovering that the euphonies, or concords, of Greek music were associated with particular ratios of lengths, especially the octave (1:2), the fifth (3:2), and the fourth (4:3). Moreover, in Greek music, these three primary concords provided the basic elements out of which *any* musical scale or composition was built. To perceive that these generating concords were associated with simple numerical ratios (in this case, of *lengths*) must have been a mind-boggling discovery in itself. For what can be more qualitative than a musical pitch or sound-quality? And what can be more subjective than a euphony?

What we can do with tone and sound, we can surely do elsewhere. This insight provided the springboard from *harmonia* to *kosmos,* the elevation of what was true in music to something encompassing the entire universe, and with it, the ineffable role of number, of measure, of value, as its manifestation. As an integrative world-view, this Pythagorean conception, intertwining as it does the good, the true, and the beautiful, the world of perceptual qualities and the world of number, still manifests its grandeur. Especially so in our world of grubbing empiricists and niggling disputators, in whose efforts we find no such qualities.

Indeed, a great deal of the Pythagorean vision clearly survives in what we call modern science, filtered in various ways through the years, and through people such as Pascal, even Gauss in later years, through Plato and Aristotle and many others in ancient times, to the latest unification schemes in theoretical physics. It is not surprising that Pythagoras became a legendary figure, even in his own time.

Commensurability

The Pythagorean story is an outgrowth of the eternal interplay between extension and enumeration, between quality and quantity, between geometry and arithmetic.

Length is one of the most immediate qualities of a geometric line segment. Measuring it, however, is a matter of enumeration, of counting. Associating a number, a length, with a particular extension or line segment is one of the most primitive examples of an *effective* process.

The first step of an effective process is to pick an arbitrary line segment A, which will serve as our unit of length and is hence arbitrarily assigned the number 1. The number assigned to any other line segment B we wish to measure has to do with how many times we can lay the unit A on the segment B; it is thus a matter of mere counting.

At this point, the Pythagoreans made an assumption: that *any* two line segments, say A and B, are *commensurable.* This amounted to supposing that, given A and B, there was another line segment U, which could be "counted off" an *integral* number of times on both A and B. This is, of course, a kind of quantization hypothesis, with U as a quantum, and as we shall see, it was wrong.

But if this is so, then we can write (committing a forgivable *abus de langage* in the process),

$$A = mU, \quad B = nU,$$

where m, n are now *integers*. Then we can further write,

$$A/B = m/n,$$

independent of U. Hence finally,

$$B = (n/m)A,$$

and hence the *rational* number (n/m) is the length of the segment B, as measured in units of A.

According to these ideas, the scheme (or software) I have outlined serves either to *measure B* given A, or to *construct B* given A. Herein lie the seeds of another fateful confusion.

But for the moment, I will restrict myself to observing that commensurability implies that, whatever the unit A, any other length B will be a *rational* number relative to it. That is, it will be the ratio of two integers.

To Pythagoras, who associated lengths with qualities such as musical tones and pitches, and ratios of integral lengths with subjectivities such as musical euphonies, such arguments must have been profoundly satisfying. Indeed, once we have posited commensurability, all this forms an incredibly tidy, closed, consistent, self-contained system, from which there is no exit *from within*. But on the other hand, there is also no way from within it to *test* the basic assumption of commensurability, no way to violate the basic link it establishes between counting and extension. And moreover, it is precisely on this notion that the effectiveness of the entire procedure rests.

"Everything Is Number"

Aristotle referred to "the Pythagoreans" often in the course of his extant writings, although his book devoted to them has been lost. In general, he was rather disapproving of them; he felt that they did not properly understand causality (and I agree). He credited them with putting forth the doctrine that "things are numbers."

Aristotle was not quite sure what was meant by this doctrine, and he gave three different interpretations of it. The first, most literally, was that real things were actually *composed* of numbers, in much the same sense that the Atomists later claimed that things were composed of atoms. The second, that things take their being from the "imitation" (*mimesis*) of numbers. And finally, that the "elements" of number, from which numbers themselves arise, were simultaneously the elements of everything.

These three statements of the Pythagorean doctrine are not necessarily inconsistent with one another and may have appeared much less so in ancient times. The following quotation from Guthrie (1962) is instructive:

> The fact that Aristotle was able to equate Pythagorean *mimesis* with Plato's notion of physical objects as "sharing in" the Ideas (which Plato himself elsewhere describes as "patterns" for the world of sense) should put us on our guard against the simple translation "imitation." The fact is, of course, that even Plato, and still more the Pythagoreans, were struggling to express new and difficult conceptions within the compass of an inadequate language. We may take a hint first from K. Joel . . . who points to the trouble that the Pythagoreans must have experienced in clearly differentiating the concepts of similarity and identity, "a defect which still plagued Sophistic thought and which Plato's Ideal theory and Aristotle's logic only overcame with difficulty because it is rooted deep in the Greek consciousness: even their language has only one word for 'same' and 'similar' (*homoios*). . . . Are things imitations of numbers or numbers themselves? Aristotle ascribes both views to the Pythagoreans, and whoever is alive to the mind of Greece will also credit them with both and agree that for them numbers served alike as real and as ideal principles."   (p. 230)

It is in fact astonishing that so much of this still lives on with us today.

The concept that "everything is number" in fact lies directly behind the attempt to subordinate geometry to arithmetic, and quality to quantity, which we have been discussing. If everything is number, then in particular the extended geometric line segments, the things endowed with length, and measured as we have described, must also be number as well. Let us pause now to express where 2,500 years have taken us in this direction; our modern language is still inadequate, but it is better than it was, and I will use it.

We start with a given set of line segments: call it L. We also presume an effective process of measurement, based on commensurability (a hypothesis about L), whereby we associate with each segment in L a (rational) number, its length relative to a preselected unit A. That is, we presuppose a mapping,

$$\lambda_A : L \to \mathfrak{R},$$

which is also (in the modern sense) effective. The (rational) numbers $\mathfrak{R}$ show up here as the range of this mapping; they comprise the set of values that length may assume, in terms of a process of counting.

But if "everything is number," we must extract these values of length that are only the range of λ_A and put them somehow into the domain; we must get them into L itself. This is in fact a very Gödelian thing to try to do, and we do it by trying to regard a line segment, an element of L, as made up out of the *values* of lengths. In particular, we go from $\mathfrak{R}$ to $2^{\mathfrak{R}}$, the set of all subsets of $\mathfrak{R}$, and try to find an image, or model, of L in there. Thus in effect we rebuild the domain of λ_A out of its range. If we can do this, we can simply throw L away and quite correctly assert that "everything *is* number" of what is left.

This is embodied in the diagram shown in figure 4.1. Here, F is part of a functor, the one that associates with every set its power set. The identification Φ is a code, or a dictionary, relating L and $2^{\mathfrak{R}}$ by identifying a line segment $\theta \in L$ with an interval of *values* of length in $\mathfrak{R}$; that is, with a set of the form

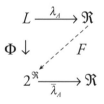

FIGURE 4.1

$$I_{ab} = \{r : a \leq r \leq b\}.$$

Now, the length $\lambda_A(\theta)$ of a segment $\theta \in L$ relative to a unit A rests on a process of counting (i.e., of counting off A on B), and the answer takes the form of an integer plus a fractional part. In $2^{\Re}$, on the other hand, we do not have to count any more. Indeed, if $\Phi(\theta) = I_{ab}$, and $\Phi(A) = I_{01}$, we can replace the original λ_A, defined on L, with a new map defined on $2^{\Re}$, where

$$\overline{\lambda}(I_{ab}) = b - a.$$

This is certainly effective, in the new universe, where "everything" is indeed number.

Notice that the new map $\overline{\lambda}$ behaves nicely with respect to set-theoretic operations in $2^{\Re}$. In particular, it is finitely additive on disjoint unions of intervals: if I_i, $i = 1, \ldots, N$ are disjoint intervals, then

$$\overline{\lambda}\left(\bigcup_{i=1}^{N} I_i\right) = \sum_{i=1}^{N} \overline{\lambda}(I_i) \tag{4.1}$$

i.e., $\overline{\lambda}$ is *finitely additive.* And this much good behavior seems to be good enough.

The trouble is that there are more things in L than can be accommodated in these diagrams. The first trouble will be that the set of *values* we are using, the rational numbers $\Re$, is in some sense too small. That will give trouble with the *range* of λ_A. If we try to enlarge that range, we will find trouble with the power set we are using to "arithmetize" L (i.e., trouble with the *domain* of $\overline{\lambda}$). In fact, these innocent-looking diagrams are going to get very muddy indeed.

Incommensurability

In light of the preceding, it was a supreme and doubtless most painful irony that it fell to Pythagoras himself to discover the falsity of commensurability. This discovery was a corollary of his most celebrated mathematical result, the one always called, since his time, the Pythagorean Theorem. And it came about by departing from the province of a single geometrical dimension, by leaving the one-dimensional world of line-segments and going to the geometric plane.

In the two-dimensional world of the plane, ratios of lengths of line segments now become associated with measuring a new kind of thing, a new kind of extension or quality called angle. The Pythagorean Theorem dealt with the angles in a right triangle. And it was an immediate corollary of that theorem that, in an isosceles right triangle, the hypotenuse *could not* be commensurable with the side. Stated another way, the sine of 45° is not a rational number.

In two dimensions, the construction of such incommensurable line segments was as effective as anything. In *one* dimension, however, it was quite impossible. Clearly, as a line segment, the hypotenuse of an isosceles right triangle had to have a length, and the Pythagorean Theorem even told us what that length must be. But how was it to be measured, as a length in itself? How could the process of its measurement be related to counting? How could we then reinterpret that measurement process as an alternate means for constructing it?

The discovery of incommensurability thus posed a number of lethal problems. It drove a wedge between the notions of measurement and construction, hitherto assumed identical, because it produced something that could be *effectively* constructed but not measured thereby. Further, it separated the notion of measurement from the notion of counting itself, which had hitherto been its basis and foundation. What, indeed, becomes of the proposition that everything is number, embodied in the tidy diagrams of the preceding section?

It was precisely at this point that mistakes began to be made, or more accurately, wrong choices taken, which have haunted us ever since. The decision made at that time was to try to *enlarge arithmetic*. In a nutshell, it was decided that we needed more numbers—irrational numbers, things to be the values of lengths such as the hypotenuse of an isosceles right triangle. That is, we must enlarge the range of the mappings λ. But that is not enough. We must also relate these new numbers back to counting, so that we may effectively measure and construct segments with these new irrational lengths. That is, we must find a way to reidentify measurement with construction in this new, larger realm.

The trouble with this is that such a program requires excursions into the realm of the arithmetic infinite. We suddenly have to count infinite numbers of things, add infinitely many summands together. The Greeks traditionally hated to do this, but at the time it must have seemed the lesser evil. However, once embarked on this path, there turned out to be (pardon the pun) no limit.

The Paradoxes of Zeno

The first hints of trouble from these procedures were not long in coming. Indeed, within a century, Zeno of Elea, a disciple of Parmenides, had put forward his devastating paradoxes. (Parmenides, by the way, argued that there was no such thing as becoming, that change and motion were illusion, and hence that there was only being.)

The problem faced by the Pythagoreans was to enlarge the rational numbers $\Re$, the set of values of λ, by including "irrationalities" such as $\sqrt{2}$. Only in this way could line segments such as the hypotenuse of an isosceles right triangle be assigned a length at all. Plane geometry gave us effective procedures for producing line segments that had to have such irrational lengths (and in fact for producing them in the greatest profusion); what was lacking was a corresponding *arithmetic* procedure for measuring and constructing them the way rational lengths could be.

The answer was to stray from the finite and admit *infinitely* many repetitions of precisely the same procedures that had worked for rational lengths. At each finite step, only rationalities would be involved, and only at the end, in the *limit,* would we actually meet our new irrational. This was the method of exhaustion, based on the observations that

$$1 \; < \; \sqrt{2} \; < \; 2,$$

$$1.4 \; < \; \sqrt{2} \; < \; 1.5,$$

$$1.41 \; < \; \sqrt{2} \; < \; 1.42, \qquad \text{and so on.}$$

In modern language, we are seeking to partition a line segment of irrational length, such as the hypotenuse θ of a right triangle of unit sides, into a *countable* union

$$\bigcup_{i=1}^{\infty} \theta_i$$

of disjoint subintervals θ_i, each of *rational* length, in a particular way, and then adding up all these rational lengths. In the terminology of equation 4.1, we are seeking to write

$$\lambda(\theta) \ = \ \lambda(\overset{\infty}{\underset{i=1}{\cup}} \theta_i) = \ \sum_{i=1}^{\infty} (\theta_i) \qquad (4.2)$$

That is, we must now require that λ be *countably additive,* not just finitely additive, as sufficed before. Under these conditions, we could write, for example,

$$\lambda(\theta) \ = \ \sqrt{2} \ = \ 1 + \ .4 + \ .01 + \ .004 + \ldots \qquad (4.3)$$

What Zeno pointed out is that this same line segment θ, via the identification in equation 4.3, is also *the countable union of its constituent points.* And each of *them* is of length *zero.* Hence, repeating precisely the above argument under these conditions leads to the paradoxical conclusion

$$\sqrt{2} \ = \ 0.$$

In fact, I could make the "length" of θ be anything I want (i.e., assume *any* value between 0 and $\sqrt{2}$) depending on how I partition it into disjoint subsets θ_i, $i = 1, 2, \ldots, \infty$.

But this is a truly devastating blow. The essence of length of a line segment is that it inheres in, and depends *only* on, the segment, not on how that segment is measured or constructed. It must in fact be independent of these; it should not depend on the specific process by which it is evaluated. What the Zeno paradoxes do, at root, is to call into question the very objectivity of the entire picture, by making the *value* of a length be determined by something extrinsic, by a particular context in which that value was determined. The problem raised by these paradoxes is thus closely akin to such things as "the measurement problem," which underlies the quantum physics of 2,500 years later, in which the observer gets causally entangled in the results of his own measurements.

Notice that these paradoxes rest upon only the *countability* of the set of *values* of length. In particular, the same argument would hold if we enlarged this set of values from the rationals $\Re$ to any set of computable numbers, numbers that arise from finite or countable repetitions of rote arithmetic operations (i.e., from algorithms) starting from integers.

The devastating paradoxes of Zeno could be interpreted in two ways, however. The first, argued tacitly by Zeno himself, is that the entire Pythagorean program to maintain the primacy of arithmetic over geometry (i.e., the identification of effectiveness with computation) and the identification of measurement with construction is inherently flawed and must be abandoned. That is, there are procedures that are perfectly effective but that cannot be assigned a computational counterpart. In effect, he argued that what we today call Church's Thesis must be abandoned, and, accordingly, that the concepts of measurement and construction with which we began were inherently far too narrow and must be extended beyond any form of arithmetic or counting.

The other interpretation, a much more conservative and tactical one, is that the thesis is sound, but we simply have not yet carried it out properly. In this light, the paradoxes merely tell us that we *still need more points,* more values for λ, that we have not yet sufficiently enlarged arithmetic. So let us add more points and see what happens.

Nonmeasurability

The ultimate enlargement of arithmetic, the transition from the original rational numbers $\mathfrak{R}$ to the full real number system $\mathbb{R}$, was not in fact achieved until the latter part of the nineteenth century, in the work of Dedekind and Weierstrass, among others. But long before then, concepts such as length, and higher-dimensional counterparts such as area and volume, had been vastly generalized, transmuted to analytic domains involving integration and associated concepts, soon to be collected into a distinguishable part of mathematics today known as Measure Theory.

In the context I have developed here, the transition from rationals to reals gets around the original Zeno paradoxes, mainly because the reals are *uncountable,* shocking as this word might be to Pythagoras. In particular, a line segment is no longer a countable union of its constituent points, and hence, as long as we demand only *countable* additivity of our measure λ, we seem to be safe. We are certainly so as long as we do not stray too far from line segments (or intervals); i.e., we limit ourselves to a universe populated only by things built from intervals by at most countable applications of set-theoretic operations (unions, intersections, complementations). This is the universe of the *Borel sets.*

But if ℝ is now the *real* numbers, figure 4.1 tells us that we are now playing with $2^ℝ$. Thus we would recapture the Pythagorean ideals, and fully embody the primacy of arithmetic expressed in figure 4.1, if it were the case that everything in $2^ℝ$ were a Borel set. We could then say that we had enlarged arithmetic "enough" to be able to stop.

Unfortunately, that is not the way it is. Not only is there a set in $2^ℝ$ that is not a Borel set, but it turns out that there are *many* of them; it turns out further that there are *so* many that a Borel set is of the greatest rarity among them. It is in fact *nongeneric* for something in $2^ℝ$ to be Borel. We shall discuss the significance of these facts, and some of the ironies embodied in them, in a moment.

To construct a set in $2^ℝ$ that is not a Borel set (or to "exhibit" one) means that we have to get far away from intervals and their syntactic, set-theoretic progeny. Let us then briefly describe how such a set can be built. It is in fact quite simple. We start with the integers Z, and a single irrational number, say our old friend $\sqrt{2}$. We form the set G of all (real) numbers of the form

$$m \div n\sqrt{2},$$

where m, n are integers. Then it is easy to show that this countable set G is everywhere dense on the line.

We will say that two *real* numbers are *equivalent* if their difference lies in G (i.e., if they are congruent *modulo G*). Again, it is easy to verify that this notion of congruence is a true equivalence relation on the real numbers ℝ, and hence it partitions ℝ into a family of disjoint equivalence classes whose union is ℝ. Since each equivalence *class* is countable (being in fact an orbit of the countable group G, of the form $r + G$, where r is a given real number), there must be *uncountably* many such equivalence classes.

Picking an element out of each equivalence class, we form a set $E \subset$ ℝ. This set *cannot be measurable;* it can have no objective length at all. Indeed, if it did have a length, that length would have to be zero. But then we are in a Zeno-like situation, for the whole set of reals ℝ is in effect a countable union of disjoint copies of sets like E. Thus countable additivity prevails, and we come up with the familiar paradox that the length of ℝ would itself be zero.

We can see the genericity of sets like E by, in effect, bumping everything up into a higher dimension. That is what we have tacitly already done in the course of the preceding argument. We have expressed the real numbers $\mathbb{R}$, a one-dimensional thing, as a Cartesian product of two subsets—basically the countable "fiber" G and the uncountable quotient set $\mathbb{R}$ *modulo* G, the "base space." Any cross section, any set of the form E, fails to be measurable as a subset of $\mathbb{R}$.

Doing this more systematically clearly reveals the nongenericity of measurability. Following an example in Halmos (1950), we can take the unit interval $I = \{r : 0 \leq r \leq 1\}$ and form the unit square $I \times I$. For any subset $U \subset I$, we can associate with U the "cylinder" $U \times I$. Define a *measure* in $I \times I$ by saying the measure of such a cylinder is the ordinary measure of its base U. Clearly, most subsets of $I \times I$ are not cylinders and hence fail to be measurable. Hence, most sets of I itself fail to be measurable.

Prodded in large part by these considerations, people quite early began looking for a way to redefine *measure* in such a way that *every* subset was measurable. By *measure,* of course, was meant a countably additive, real-valued measure, such that individual points are of measure zero. We have just seen that ordinary (Lebesque) measure, tied ineluctably to intervals, is very far from this, and accordingly, if such a measure even exists, it must take us far from our intuitive ideas going back to Pythagoras.

Indeed, a set S that could support such a measure must, by any criterion, be a very strange set. Among other things, it must be very large indeed—large enough so that, roughly, 2^S is not much bigger than S. Such an S is called a *measurable cardinal.* The existence of such a measurable cardinal turns out to contradict the "axiom of constructibility," which is in itself a direct descendant (in fact, a close cousin) of the ancient attempt to subordinate quality to quantity manifested in what we have discussed.

In another direction, these ideas lead to such things as the Banach-Tarski paradox: a measurable set can be dissected into a small number of *non*measurable subsets (i.e., it can be expressed as their disjoint union). These nonmeasurable fragments can then be moved around by the most benign transformations (e.g., rigid motions, which preserve measure, if there is a measure to be preserved) and thereby reassembled into a different measurable set, of arbitrarily different measure than the one we started with. In a way, this is the other side of the Zeno paradox; similarly, this one exhibits the ultimate in context-dependence of length.

General Discussion

There are many ramifications indeed of the ideas we have been discussing. As we cannot pursue most of them here, we will concentrate on those connected with Church's Thesis. I shall argue that the thesis rests, at root, on an attempt to reimpose a restriction such as commensurability, and to deny existence, or reality, to whatever fails to satisfy that condition.

In an old film called *Body and Soul,* the racketeer Roberts represents the best possible expression of the thesis. He says, "Everything is either addition or subtraction. The rest is conversation."

As we have seen, commensurability was the original peg, the *harmonia,* that tied geometry to arithmetic. It said that we could express a geometric quality, such as the length of an extended line segment, by a number. Moreover, this number was obtained by repetition of a rote process, basically *counting.* Finitely many repetitions of this rote process would *exhaust* the quality. But more than this, it asserted that we could replace the original geometric *referents* of these numbers by things generated *from* the numbers themselves, and hence we could completely dispense with these referents. As I described in *Life Itself,* the elimination of (external) referents amounts to the replacement of semantics by syntax.

The repetition of rote operations is the essence of *algorithm.* The effect of commensurability was to assert an all-sufficiency of algorithm. In such an algorithmic universe, as we have seen, we could always equate quality with quantity, and construction with computation, and effectiveness with computability.

Once inside such a universe, however, we cannot get out again, because all the original external *referents* have presumably been pulled inside with us. The thesis in effect assures us that we never *need* to get outside again, that all referents have indeed been internalized in a purely syntactic form.

But commensurability was false. Therefore the Pythagorean formalizations did not in fact pull qualities inside. The successive attempts to maintain the nice *consequences* of commensurability, in the face of the falsity of commensurability itself, have led to ever-escalating troubles in the very foundations of mathematics, troubles that are nowhere near their end at the present time.

What we today call Church's Thesis began as an attempt to internalize, or formalize, the notion of effectiveness. It proceeded by *equating* effectiveness with what could be done by iterating rote processes that were already inside—i.e., with algorithms based entirely on syntax. That is exactly what *computability* means. But it *entails commensurability*. Therefore it too is false.

This is, in fact, one way to interpret the Gödel Incompleteness Theorem. It shows the inadequacy of repetitions of rote processes in general. In particular, it shows the inadequacy of the rote metaprocess of adding more rote processes to what is already inside.

Formalization procedures, such as the one contemplated by Pythagoras so long ago and relentlessly pursued since then, create worlds without external referents; from within such a world, one cannot entail the *existence* of anything outside. The idea behind seeking such a formalized universe is that, if it is big enough, then everything originally outside will, in some sense, have an exact image inside. Once inside, we can claim objectivity; we can claim independence of any external context, because there is no external context anymore. Inside such a world, we can literally claim that "everything is number" in the strongest sense, in the sense of *identity*.

But if we retain an exterior perspective, what goes on within any such formalized universe is at best *mimesis* of what is outside. It is simulation. Hence, if we cannot eliminate the outside completely, cannot claim to have fully internalized it, the next best thing is to claim that whatever is not thus internalized can be simulated by what has been, that is, simulated by the repetition of rote processes (algorithms), and rest content with these approximations.

But once we have admitted an outside at all, we can no longer equate effectiveness and computation, or equate construction and measurement, or in general equate *anything* that happens outside with the algorithms inside. Mimesis does not extend that far; the geometric construction of $\sqrt{2}$ from the Pythagorean Theorem has nothing to do with the arithmetic algorithm that computes it, or measures it, or "constructs" it. In short, mimesis is not science, it only mimics science.

Throughout the foregoing discussion, and indeed throughout the history of mathematics, we have found ourselves being relentlessly pushed from small and tidy systems to large ones. We have been pushed from integers to rationals to reals; from sets of elements to sets of sets. We have repeatedly tried to pull the bigger sets back into the smaller ones. We keep finding that they do not generally fit.

Perhaps the main reason they do not fit is that, while we are (grudgingly) willing to enlarge the sets of *elements* we deal with, we are unwilling to correspondingly enlarge the class of *operations* through which we allow ourselves to manipulate them. In particular, we have found ourselves repeatedly going from a set S to a power set 2^S. But operations on S become highly nongeneric in 2^S. Or stated another way, we cannot expect to do much in 2^S if we only permit ourselves those operations inherited from S. We do not have enough *entailment* to deal with 2^S properly under these circumstances.

Klein's famous Erlangen Programm in geometry gives an example. In his efforts to redefine geometry, Klein posited that it is the study of pairs (S, G), where S is an arbitrary set of *points,* and G is a *transitive* group of transformations (permutations) of S. But the set of geometric *figures* is the set of subsets, 2^S, a much bigger set. We can make the original group G operate on 2^S in the obvious way, but on 2^S, G has become highly *nontransitive.* Whereas G had only one orbit (S itself) while operating on S, it now has many orbits on 2^S; these are the congruence classes. Geometry is hard, in this light, precisely because the operations in G are rare, considered in the context of 2^S, whereas they were not rare in the original context of S. And, of course, we can iterate the procedure indefinitely, in the course of which the operations from G become rarer and rarer.

Computability is just like this. That is precisely why, as I have said in *Life Itself,* formalizable systems are so incredibly feeble in entailment. They attempt to do mathematics in large realms with only those procedures appropriate to small ones. As the realms themselves enlarge, those procedures get more and more special and, at the same time, less and less objective. That is why formalizable systems themselves are so nongeneric among mathematical systems.

From Pythagoras to Church, the attempt to preserve smallness in large worlds has led to a succession of disastrous paradoxes, and it is past time to forget about it. At the outset, I said that the Pythagorean program had seeped insidiously into epistemology, into how we view science and the scientific enterprise, through common concerns with measurement and the search for laws relating them. Pythagorean ideas are found at the root of the program of reduction, the attempt to devolve what is large exclusively on what is small and, conversely, to generate everything large by syntactic processes entirely out of what is small. Another manifestation is the confusion of science with mimesis, through the idea that whatever is, is simulable. This is embodied in the idea that *every model of a material process must be formalizable.* I have discussed the ramifications of these ideas, and especially their impacts on biology, extensively in *Life Itself.*

In particular, these ideas have become confused with *objectivity,* and hence with the very fabric of the scientific enterprise. Objectivity is supposed to mean observer independence, and more generally, context independence. Over the course of time, this has come to mean only building from the smaller to the larger, and reducing the larger to the smaller. As we have seen, this can be true only in a very small world. In any large world, such as the one we inhabit, this kind of identification is in fact a mutilation, and it serves only to estrange most of what is interesting from the realm of science itself.

Epilog: A Word about Fractals

For the Greeks, the length of a curve, like an arc of a circle, was the length of a line segment obtained when the curve was straightened out. It was presumed that this process of straightening could always be performed without changing that length. This is another facet of commensurability.

An integration process, such as a line integral, basically embodies such a straightening process, approximating the curve by segments of smaller and smaller lengths. These approximations distort the curve, but in the limit, when the unit is infinitely small, the curve is presumably effectively straightened without distortion, and the length of the resulting line segment is assigned to the original curve.

Ideas of this kind underlay Mandelbrot's (1977) original conception of fractals. From this point of view, a fractal is a curve that cannot be straightened out into a line segment without distortion, without stretching some parts of it and compressing others. If we try to measure the length of such a thing by any process of unrolling, the number we get will depend entirely on how we straightened it, not on the curve itself.

Indeed, it is in some sense nongeneric for a curve to be thus straightened without distorting it. Stated another way, there are no algorithms for straightening arbitrary curves of this type, no algorithms for computing their lengths. We can, in a sense, effectively generate (some of) them, albeit via an infinite process. But their *analysis* immediately violates Church's Thesis. Zeno would have loved them.

❧

CHAPTER 5

Drawing the Boundary Between Subject and Object: Comments on the Mind-Brain Problem*

The discussion to follow will focus on the obvious irreconcilability of the two statements below. The first of them is taken from a rhapsodic paean to reductionism; it was penned over two decades ago, but could have been written yesterday:

> Life can be understood in terms of the laws that govern and the phenomena that characterize the inanimate, physical universe, and indeed, at its essence, life can be understood *only* in the language of chemistry. . . . Only two truly major questions remain shrouded in a cloak of *not quite* fathomable mystery: (1) the origin of life . . ., and (2) the mind-body problem; i.e. the physical basis for self-awareness and personality. (Handler 1970; *emphasis added*)

The second was written by a physicist, and expresses his view of what physics is about:

> [Max] Planck designated in an excellent way . . . the goal of physics as the complete separation of the world from the individuality of the structuring mind; i.e., the emancipation of anthropomorphic elements. That means: it is the task of physics to build a world which is foreign to consciousness and in which consciousness is obliterated. (Bergman 1973)

---

*This chapter was previously published in the journal *Theoretical Medicine* 14 (1993): 89–100; copyright © 1993 by Kluwer Academic Publishers. Reprinted by permission.

Here we see, in a nutshell, the roots of the real problem. Handler claims to be seeking the physical, material basis of things (life and mind) which physics has extruded. According to Bergman, neither life nor mind can be either subject or object in a true science of material nature, of *objective* reality. At best, according to this view of *physics,* life and mind are only epiphenomena; *façons de parler;* mere names given to certain classes of objective, material events and behaviors; devoid of any causal agency.

As I shall argue below, the real issue here is not a technical, but a conceptual matter. The central concept at issue is *objectivity,* for this alone determines whether something falls into the realm of science or not. As it stands now, we must say that organisms are objective but life is not, brains are objective but mind is not. Accordingly there can be a science of organisms, and a science of brains, but no science of life or of mind. Conversely, if life and mind are to be made objects of *scientific* study, it is our conceptions of objectivity which must change.

The identification of *objectivity* with what is independent of or invariant to (these are not the same) perceivers, or cognizers, or observers, is what has led to the current infatuation with the *machine* as simulacrum of both life and mind. Roughly speaking, if a machine can "do" something, that is prima facie evidence of its objectivity, and of its admissibility into science. Hence the conflation of mechanism with what is objective, and the relegation of anything nonmechanistic to the realm of the subjective, the ad hoc, the vitalistic, the anthropomorphic: in short, beyond the pale of science.

I shall argue, to the contrary, that mechanism in any sense is an inadequate criterion for objectivity; that something can be objective (and hence a candidate for scientific scrutiny) without being in any sense a mechanism. That is, the perceived dichotomy between mechanism and vitalism (i.e., denial of the former means affirmation of the latter) is a false one.

Under these circumstances, we shall argue that Handler's "not quite fathomable" problems are so because of a mistaken equation of mechanism with objectivity; accordingly, it will also follow that the physics Bergman describes, in which "consciousness is obliterated," is only a small part of a full science of material nature or objective reality.

Objectivity

The world "foreign to consciousness," which Bergman evokes in the above citation, comprises the external world of events and phenomena: the *public* world. The world of mind and consciousness, on the other hand, is a private world; what happens in it, and in whatever it touches, constitutes the *subjective*. We should notice that such a partition of our immediate universe of impressions and percepts, into a public part and a private part, is itself private. There is no public test for "publicity"; it is something privately posited.

The entire concern of science, especially theoretical science, has been to make public phenomena apprehensible to the private, cognizing mind. The mind-brain problem represents an attempt to go the other way: to pull the private world into the public one, and thus to make the mind apprehensible to itself by expressing it in phenomenal terms.

In one way or another, the concept of *causality* dominates our conception of what transpires in the external, public world. In broadest terms, causality comprehends a system of entailments, which relate the events and phenomena occurring therein. The concept itself is due to Aristotle, who associated it with the answers given to the question, Why?; indeed, to Aristotle, science itself was the systematic study of "the why of things," and hence entirely concerned with elucidating such causal relations.

Intuitively, we think of something as "objective" if its perception, or cognition, plays no causal role in its *entailment;* i.e., answers no question *why* about it. Stated another way: the private perceptions or cognitions about the thing have no public counterparts or manifestations in the thing itself. It is only in this sense that "objective" things are foreign to consciousness.

In practice, objectivity has become more narrowly construed still, especially in biology. It has come to mean not only independence from a perceiving mind but independence from any larger system whatever, public or private, of which it may (or may not) be a part: hence, independence from any *environment.* Accordingly, the notion of function, which depends on just such a larger system, is disallowed any objective status, likewise any notion of emergence. This is one of the wellsprings of reductionism: that it is only objective to explain wholes in terms of parts, never parts in terms of wholes.

So, for example, a molecule of DNA, or even just its *sequence,* are accorded objective status; they can be removed from their original environment without affecting these characteristics, and hence we can ask why about them without ever invoking the organism from which they came. The Mendelian gene, on the other hand, had no such status, since it had to be defined exclusively in *functional* terms, through manifestations in phenotype. On similar grounds, the Behaviorist school in psychology has stoutly denied "objectivity" to what it calls *internal mental states*, reserving that term entirely for functional input-output (stimulus-response) characteristics analogous to those of classical genetics.

Perhaps the strongest statement about objectivity was given by Jacques Monod:

> The cornerstone of the scientific method is the postulate that nature is objective. In other words, the systematic denial that "true" knowledge can be gotten at by interpreting phenomena in terms of final causes . . . the postulate of objectivity is consubstantial with science . . . there is no way to be rid of it . . . without departing from the domain of science itself.   (1971:21)

We shall come to what "final causes" means shortly.

The upshot of all of this is the following. First, we must allow public events to have only public explanations, public causes, never private ones. Second, in that public arena, causal chains must always flow from parts to wholes, never from wholes to parts. This, and only this, constitutes the domain of "objective reality."

In Aristotelian terms, then, objectivity means in practice an exclusive concern with what can be accommodated entirely within *only three* of his four posited causal categories. Notions of material, formal, and efficient causation alone are necessary and sufficient for the external, public world; final causation is excluded, and relegated at best to the private, subjective one. Accordingly, the world of the objective allows its events to be displayed along one single, coherent time-frame ("real" time), in which causal entailment flows from past through present to future, and never the other way. Thus arises what I have called the Zeroth Commandment: Thou shalt not allow future state to affect present change of state. Systems in which this happens are called *anticipatory* (see my *Anticipatory Systems*) and accordingly are dismissed as "acausal."

The main consequence of these views of objectivity is that *closed causal loops are forbidden.* Only chains or trees are admitted into the objective world, arrayed like branches along the trunk of that single, coherent, all-encompassing time-frame.

Our argument will be that, if closed loops of causation are denied objective status per se, then the mind-brain problem falls irretrievably outside of science. Systems without closed causal loops are, broadly, what I have called *machines* or *mechanisms,* or more generally, *simple systems* (see *Life Itself*). What makes simple systems simple in this sense is, roughly, that they are so weak in entailment that there is not enough to close such a causal loop. As we shall see, the alternative is not a "subjective" world, immune to science, but a world of *complex systems,* i.e., of systems which are not simple.

Causality and "The Measurement Problem"

As we have used it above, the term *causality* pertains to entailment relations which can be established between public events, especially between events occurring "here and now," and other events occurring "there and later (or earlier)." In physics, this causality has always been ineluctably connected with the notion of *state* (or, more precisely, with the state of a "system" at a particular instant). Intuitively, the information comprising such a state consists of all you need to know to answer every question why about the system's behavior. It also connotes a notion of parsimony: that a state is the least you need to know to answer these questions. The determination of state, however, requires *observation;* it requires measurement.

In these terms, the partition between what is objective and what is not transmutes into the partition between a "system" and an *observer.* That partition has been most extensively and most urgently discussed in the context of quantum theory, and that from its very earliest days. That continuing discussion has been, to say the least, inconclusive. But several things have emerged from it, which bear on our present considerations: (1) *where* the partition between "system" and observer is drawn is entirely arbitrary; (2) *wherever* the partition is drawn, it always seems to leave some "physics" (i.e., something public or objective) on the wrong side of it; and (3) the more we admit into the objective or public side, the more porous the partition itself seems to become.

Concern with "the measurement problem" was, for instance, a central theme of von Neumann's (1955) early monograph on quantum theory, so much so that at least a third of its pages are devoted to it. The root of the problem, of course, lay in the fact that, at the quantum level, observation apparently had to be invoked to account for the result of observation. This immediately flies in the face of objectivity itself, where as we have seen, something is to be counted as objective only if it is (causally) independent of observation; if observing something answers no question about why it is what it is. At the very least, measurement appears to invoke a causal flow from a whole (system + observer) to a part (system alone) which, as we noted earlier, is stoutly denied "objective" status.

Let us briefly consider von Neumann's treatment of these matters, because they are directly pertinent, and because they have provided the basis for many subsequent discussions. First, as to the necessity to make *some* partition between observer and the observed, between the objective and the subjective:

> It is inherently entirely correct that the measurement or the related process of the subjective perception is a new entity relative to the physical environment and is not reducible to the latter. Indeed, subjective perception leads us into the intellectual inner life of the individual, which is extra-observational *by its very nature.*   (1955:418; *emphasis added*)

Then he goes on to say

> We *must* always divide the world into two parts, the one being the observed system, the other the observer. The boundary between the two is arbitrary to a very large extent . . . this boundary can be pushed arbitrarily deeply into the interior of the body of the actual observer . . . but this does not change the fact that in each method of description the boundary *must* be put somewhere, if the method is not to proceed vacuously.   (1955:420)

Von Neumann, following Bohr in this, invokes the necessity for such a partition to justify a dualism mandated by quantum theory itself to deal with the "measurement process." Namely, a quantum-theoretic system will behave causally as long as it is not being observed, but "non-causally" (i.e., statistically) otherwise. The causal aspect is embodied in, say, the system's own Schrödinger equation governing autonomous change of state. But this equation no longer holds when change of state arises from non-autonomous interactions, e.g., with measuring instruments. These latter create dispersions, irreversibilities, and dissipations, which have no classical counterparts, and which make the "state" of the system a much more problematic concept, in terms of its causal content (i.e., the why questions it can answer) than it was previously.

Actually, von Neumann's primary concern is to affirm quantum theory by denying causality, relegating it to the status of a simple macroscopic illusion:

> The position of causality in modern physics can therefore be characterized as follows: In the macroscopic case there is no *experiment* which supports it, and none can be devised because the apparent causal order of the world in the large . . . has *certainly* no other cause than the "law of large numbers" and it is completely independent or whether the natural laws governing the elementary processes are causal or not. . . . The question of causality could be put to a true test *only* in the atom, in the elementary processes themselves, and here everything in the present state of our knowledge militates against it. . . . We may [still] say that there is at present no occasion and no reason to speak of causality in nature—because no experiment indicates its presence, since the macroscopic are unsuitable in principle, and . . . quantum mechanics contradicts it.   (1955:326–327)

As we have seen, the cost of these sweeping assertions is to force the observer (i.e., the subjective) to intrude into the objective world, and indeed, in a way which makes the intrusion itself objective. We are thus mandated to do precisely the thing which objectivity denies, namely, to involve the subjective observer in accounting for what is observed. That is, the boundary between subjective and objective has become porous. And it is equally clear that the more we try to include on the objective side of our boundary (i.e., the more microscopic a view we take, and the more of "the body of the observer" we include in it), the more porous it gets, the less of a separation it actually makes.

The real problem lies in trying to tie objectivity irrevocably to a putative notion of state, and thereby restricting causality to pertain *only* to

(a) the determination of observables, and (b) the determination of state transitions, by state. In fact, causality, in the original Aristotelian sense, means much more than this. Accordingly, we shall argue that any science which mandates this, including quantum theory, must be a very special science; in particular, it will have its own version of the measurement problem or of the life-organism problem, or the mind-brain problem.

Of course, in the above few pages we can hardly do justice to the ramifications of the measurement problem (for a survey, see Wheeler and Aurek 1983).

Cognate Anomalies in Mathematics

The domain of mathematics lies entirely within the inner, private, subjective world; ironically, however, that domain is also considered the most objective of realms. From at least the time of Pythagoras, "mathematical truth" was the best truth, independent not only of the mathematician but of the external world itself. A culmination of this was the development of Platonic Idealism, in which a "real thing" is regarded as something mathematical (its Idea) plus the corruption occasioned by attaching that Idea to a specific (hence imperfect) external referent. Materialism, in any of its many forms, can likewise be regarded as an opposite attempt to pull this mathematical truth into conventional perceptive realms arising in the external world. We can see in this opposition the germs of the mind-brain problem itself.

In any event, the mathematical universe comprises systems of entailment (inferential entailment) no less compelling than the causal relations governing objective events in the external world. Indeed, inferential entailment (between propositions) and causal entailment (between external events) are the only two modes of entailment we know about. We can in fact deal with both of them in exactly parallel Aristotelian terms, by asking why.

The surprising fact is that these two different realms of entailment (the objective world of causal entailment and the relatively subjective realm of inferential entailment) run so much in parallel. The physicist Wigner (1960) thought of this as *unreasonable,* and it is still a matter of lively debate (Mickens 1990). In fact, the congruences (modelling relations) which can be established between them are, I would argue, the essential stuff of science (see my *Anticipatory Systems*).

Over the past century, the mathematical realm has run into a great deal of trouble, a foundations crisis, just as physics has. This current crisis was touched off by the appearance of paradoxes, some of which (e.g., the "liar paradox" of Epimenides) actually go back to ancient times.

Most, if not all, the known paradoxes arise from an attempt to divide a universe into two parts on the basis of satisfying some property or not (e.g., a property like objectivity). *Trouble arises whenever this property can be turned back on itself,* in particular, when we try to put some consequent of the property back into one or the other class defined *by* the property. This constitutes an *impredicativity,* what Bertrand Russell called a *vicious circle.* Kleene puts the matter as follows:

> When a set M and a particular object m are so defined that on the one hand m is a member of M, and on the other hand the definition of m depends on M, we say that the procedure (or the definition of m, or the definition of M) is *impredicative.* Similarly when a property P is possessed by an object m whose definition depends on P (here M is the set of the objects which possess the property P). An impredicative definition is circular, at least on its face, as what is defined participates in its own definition. . . . In the Epimenides paradox [i.e., Epimenides the Cretan says that all Cretans are liars] the totality of statements is divided into two parts, the true and the false statement. A statement which refers to this division is reckoned as of the original totality, when we ask if it is true or false.   (1950:42)

On the other hand, as Kleene states,

> Thus it might appear that we have a sufficient solution and adequate insight into the paradoxes *except for one circumstance;* parts of mathematics we want to retain, particularly analysis, also contain impredicative definitions. An example is the definition (of the least upper bound of an arbitrary bounded set of real numbers).   (1950:42; *emphasis added*)

What has come to be called *constructible mathematics* is an attempt to eliminate all the inferential loops or "vicious circles," thereby on the one hand eliminating the basis of all the paradoxes, and on the other, providing equivalent *predicative* definitions of presently impredicative ones, like "least upper bound," (which, it may be noted, is the basis for every approach to *optimality,* among many other things). It was widely supposed that *only* the things in constructive mathematics *had any objective basis for existing.*

Russell's "Theory of Types" represents one kind of attempt to straighten out all the impredicative loops of inferential entailment in mathematics. It was a failure, which lost itself in at least equally bad infinite regresses of unlimited complication. Another, more modern attempt in this connection relies on a "constructible universe," originally proposed by Gödel as a *model* for set theory, in connection with determining the status of such things as the Axiom of Choice and the Continuum Hypothesis.

Such a constructible universe is one which starts with a finite (usually small) number of elementary syntactic operations and a minimal class of generators for them to operate on. Time moves in discreet ordinal steps. At each stage, new things can be constructed, by applying one of the rules to what has been constructed in the preceding stages. Thus, no closed loops, no impredicativities can arise in that universe. In particular, anything in that universe has a pedigree, going back to the original elements, through specific rules applied at successive ordinal time-steps; i.e., it is *algorithmically* generated.

Without going into details, we can see already the machine-like character of this kind of constructible universe. This is precisely what is supposed to make it objective. As in the causal world, no later stage can affect *any* earlier one, certainly not the original generators of which it is comprised.

Another language to describe this kind of constructible universe is that of *formalization.* A mathematical system is formalizable if everything in it can be generated in such a constructible fashion. Formalization was Hilbert's answer to the paradoxes, and consisted in essence of stripping the system of all referents (even mathematical ones). In effect, mathematics was to become a game of pattern generation, played by applying syntactic rules governing the manipulation of the *symbols* in which the mathematics was expressed. To quote Kleene again:

> [Formalization] will not be finished until all the properties of undefined terms which matter for the deduction of theorems have been expressed by axioms. Then it should be possible to perform the deductions treating the technical terms as words in themselves without meaning. For to

say that they have meanings necessary to the deduction of the theorems, other than what they derive from the axioms which govern them, amounts to saying that not all of their properties which matter for the deductions have been expressed by axioms. When the meanings of the technical terms are thus left out of account, we have arrived at the standpoint of formal axiomatics.   (1950:59)

Where there are no referents at all, there are *a fortiori* no self-referents. Moreover, the internalizing of such referents in the form of additional syntactic rules is precisely the basis for regarding formalizations as objective. Once again, no impredicativities can arise in such a formalization. It was Hilbert's program to thus formalize all of mathematics.

The status of all these formalizations is informative. They turn out to be infinitely feeble compared with the original mathematical systems they attempted to objectivize. Indeed, these attempts to secure mathematics from paradox by invoking constructibility, or formalizability, end up by losing most of it. This is one of the upshots of Gödel's celebrated Incompleteness Theorem (Gödel 1931), which showed precisely that "self-referential" statements (e.g., "this proposition is unprovable in a given formalization"), which are perfectly acceptable in the context of ordinary Number Theory, fall outside that formalization. In other words, a "constructible universe" is at best only an infinitesimal fragment of "mathematical reality," considered as a system of inferential entailment, no matter how many elements, or how many syntactic rules, we allow into it.

This situation should remind the reader of what we have already seen. We can push the boundary between what is constructive, or formalizable, arbitrarily far into mathematics. But wherever we draw that boundary, more remains on the wrong side of it. On the formalizable (objective) side of the boundary, there are no closed loops of entailment, just as, in the causal realm, we exclude them by restricting ourselves to those causal categories (material, formal, and efficient causes) which only go in a fixed time-frame from earlier to later. Thus we can see, in particular, that there is no way to ever map the nonformalized side into the formalized one *in an entailment-preserving fashion*. There are many ways to say this: we cannot replace semantics by syntax; mathematics is much more than word processing or symbol manipulation; mathematics transcends algorithms; mathematics cannot be expressed as program to fixed, finite-state hardware.

In a certain sense, then, formalizable mathematical systems (i.e., those without impredicativities) are indeed infinitely feeble in terms of entailment. As such, they are excessively nongeneric, infinitely atypical of mathematical (inferential) systems at large, let alone "informal" things like natural languages. Any attempt to objectify all of mathematics by imposing some kind of axiom of constructibility, or by invoking Church's Thesis (Rosen 1962a, 1988a), only serves to estrange one from mathematics itself. Indeed, what is necessary is quite the opposite: *to objectify impredicative loops* (Löfgren 1968).

Implications for the Mind-Brain Problem

We will now try to pull together the various threads we have developed above, and indicate their bearing on the mind-brain problem.

From what has been said above, the "objectivizing" of the observer (i.e., pulling him entirely into the public, external world) amounts to replacing the boundary between subjective and objective by an ordinary boundary between a "system" and its environment, both now in the external world. Moreover, this must be done in such a way that what, formerly, was (subjective) inferential entailment in the observer now *coincides* with causal entailment in the objective system which has replaced him. At the very least, there must be no less *causal* entailment in that system than there was *inferential* entailment in the subjective observer.

These requirements are inconsistent with the tenets of mechanism, tenets which, as we have seen, have been presumed synonymous with objectivity itself.

For instance, *mind* requires unformalizability to be part of it. That means, precisely, the accommodation of closed loops of inferential entailment. But as we have seen, the identification of mechanism with objectivity forbids closed *causal* loops. And yet the presumed "objectivization" of the observer must faithfully represent his inferential entailments in terms of objective causal ones.

That is one thing, which by itself would be quite decisive. But there are many others. For instance, part of the subjective world must comprise the *referents* he establishes between his internal models and the systems he sees in the external world. It is clear that any attempt to "objectivize" these creates an immediate *self*-reference *in the presumed objectivization,* where no referents are allowed at all. Indeed, this situation spawns actual paradoxes, involving mappings forced to belong to their own domains and ranges (Rosen 1959).

The conclusion I draw from these circumstances is that, so long as we persist in equating mechanism with objectivity, and hence with science, the mind-brain problem and, even more, the life-organism problem are inherently outside the reach of that science. The alternatives, in the starkest possible form, are either to give up the problem or to change the science. As to the latter, the change required is clear, it is abandoning the equation of objectivity with mechanism. We must accordingly allow an objective status to complexity as defined above, i.e., to systems which can accommodate impredicativities, or closed loops of entailment.

Actually, such an enlargement seems much more radical in concept than it would be in practice. Mechanism would not thereby disappear: it becomes simply a limiting case of complexity. So, for example, the discovery of irrational numbers was a much more radical conceptual change than a practical one. Nevertheless, moving from rational numbers to real numbers was a revolutionary step, certainly in terms of the conceptual possibilities it opened up. So it is here.

Let me address one objection, which is always raised at this point. It is this. Surely, each individual particle in "the body of the observer" is a mechanical thing. And (at least at any instant) there are only a finite number of them. Hence, if we pull all these constituent particles into the "objective world," one by one, the observer himself will necessarily come with them. Hence the observer is himself mechanical, and indeed could be objectively re-created, particle by particle. In particular, we could re-create a subjective mind by thus assembling an objective brain.

There are many flaws in such an argument. The most cogent is that the putative assembly process, the process of re-creation itself, has become complex, hence non-"objective." In addition, such a process would actually obliterate the mind-brain interface (or, more generally, the subjective-objective boundary), which is now merely a system-environment boundary. Obliterating the boundary would leave us with the entire universe: either all environment and no system, or all system and no environment. And we recall that, as von Neumann (1955) argued, such a boundary must be put somewhere, "if the method is not to proceed vacuously."

It should be stressed that, by advocating the objectivity of complex systems, systems with nonformalizable models and hence closed loops of entailment (impredicativities), I am advocating the objectivity of at least a limited kind of *final* causation. This is precisely what closes the causal loops. It simply describes something in terms of what it entails, rather than exclusively in terms of what entails it. This, it will be observed, need have nothing whatever to do with *Telos*, any more than, say, Gödel's Incompleteness Theorem does.

In such a complex world, furthermore, functional descriptions are perfectly meaningful, and can be quite independent of any mechanistic ones. And, since we are freed from the exigencies of a single constructive or algorithmic time-frame, mechanistic objections to anticipation, and in particular to internal predictive self-models which provide its basis, no longer apply at all. I have dealt with all these matters at greater length in *Life Itself.*

∾

CHAPTER 6

Mind as Phenotype

Phenotype is a uniquely biological concept. Etymologically, it derives from the Greek verb *phainein,* to show, the same root from whence derives the word *phenomenon.*

Phenotype is a rather recent word, but *phenomenon* is old. The dictionary tells us that *phenomenon* means "any fact, circumstance or experience that is apparent to the senses, and that can be scientifically described or appraised." That is, phenomena devolve on *cognition,* and hence on direct (or sometimes indirect) *sensory* experience. From this follows another characterization of *phenomenon,* typically regarded as synonymous with the one I have given: the appearance or observed features of something experienced, as distinguished from reality or the thing in itself. Indeed, the profound limitations of cognition as an exclusive guide to this reality, this *ding an sich,* have been emphasized by philosophers from Plato to Kant.

In biology, *phenotype* thus pertains to what we can see or measure about an organism, or about its behaviors, or its properties—its anatomy and its physiology. It thus comprises whatever there is about organisms that directly exposes them to cognition in the first place, and that which renders such knowledge sufficiently *objective* for scientific purposes.

But in biology, it for the first time in science becomes apparent that phenotype is not enough, that the objective cognition of such visible properties does not and cannot suffice to make biology a *science.* That is why the term *phenotype* comes paired with a dual concept, *genotype,* which pertains not directly to phenomena themselves but rather to their causal bases. In other branches of science, such as standard physics and chemistry, there was deemed no need for any such distinction—genotype and phenotype simply *coalesce* in them, united in a common world of phenomena. But in biology, they do not coalesce any more; they have somehow *bifurcated*. Indeed, one of the fundamental problems of approaching even phenotypes *reductionistically,* exclusively in terms where genotype and phenotype do coincide, resides precisely here. We shall look at such questions from a variety of viewpoints.

If biology raises such questions already, they are raised even more vividly by considerations of mind and sentience or consciousness. The fact is that each of us is more aware of these than we can possibly be of any mere cognition or sensory impression. The awareness is thus *phenotype,* but it has nothing to do with cognitions of phenomena, which indeed presuppose it. Above all, it is *subjective,* and this subjectivity is intensely contagious, spreading to everything it touches—volition, intentionality, imagination, emotion, and all the rest of it, including cognition, and mathematics, and science.

Here, we shall explore two quite different ways of attempting to "objectify" the subjective phenomena of mind and sentience, of pulling them into the cognitively based world of phenomena and phenotype. First, there is a strategy I call *mimesis* (see chapter 4), based on the idea that an understanding of any property resides entirely in finding some purely mechanical device that mimics the property "sufficiently closely." We can then *impute,* from the mimicry itself, the causal basis of the behavior from the mimic to what it mimics—a tacit invocation of Occam's Razor. In particular, if we can find such a machine that mimics in its behaviors some aspect of "intelligence" or "thought," then we can say the *machine* already "thinks." Thus understanding of *thought* devolves entirely on understanding that mechanical mimic. In effect, we are turning the mimicry relation around and extrapolating from the machine to anything that *mimics the machine.* This, of course, is the substance of the familiar Turing Test (see later). We can already see, however, that arguments of this kind teeter on the brink of equivocation.

A second approach is loosely called *reductionism.* The basic idea here is to devolve any phenotypic property of a material system onto those manifested by a sufficiently extensive class of material subsystems, such as chemical molecules, atomic particles, subatomic particles, and points south. These subsystems are themselves required to be *context independent,* in the sense that (1) they must be fractionable, or separable, from any larger system to which they belong, and (2) their own properties do not depend on where they came from or how they were isolated. A phenotypic property of such a larger system is objective to the extent that it is expressible entirely in terms of those of such constituents. The approach here is thus poles apart from mimesis; instead of relating material systems through their capacity to behave alike, independent of what they are made of, we rather relate systems, however diverse in phenotype, through their reducibility into such context-independent elements. Stated another way, any phenotypic property of any material system is but another way of perceiving, or of apperceiving, the very same properties of the elements that we can cognize directly about them. In particular, a property like sentience, a subjective property of certain material systems, is but a reexpression and reformulation of those same cognitive properties of electrons, protons, neutrons, and what have you that we perceive directly about them when they are in isolation. To use a philosophic term, any phenotypic character of any material system is already *immanent* in a few properties of a few kinds of context-independent reductionistic elements. It is precisely this immanence which makes the procedure universal.

Each of these pictures of objectivity, considered by itself, seems persuasive enough. The trouble is that they do not coincide; indeed, they are not even compatible with each other. One picture (the mimetic one) pertains to phenotype alone; the other (the reductionistic one) pertains to genotypes. If they are to be coextensive, the two must indeed coincide. But in organisms, as we have seen, they do not coincide. Moreover, any world in which they do coincide (e.g., the world of *algorithm*) is a barren, sterile one, far too impoverished to do mathematics in, let alone support anything resembling life, let alone mind and sentience.

A specific consequence of the reductionistic picture I have sketched is this, that one must not expect to learn anything *new* about reductionistic units (e.g., about electrons, or protons, or neutrons) by looking at phenotypes of larger systems, like organisms. To suggest otherwise is to strike a blow at reductionism itself, and its own presumed identification with objectivity; those who believe in reductionism see it as an advocacy of *vitalism*. Be that as it may, the thesis of the following argument is that, if life and sentience are phenotypic, then *physics* is not what we presently think it is. And likewise, even if they are *not* phenotypic.

The Turing Test

The Turing Test asserts several things. First, it proposes an effective procedure for discriminating phenotypes into two classes, which we may call intelligent and unintelligent. It thus proposes this "intelligence" as itself an objective phenotypic property of something, as objective in its way as the charge on an electron. Second, it asserts the nonexistence of any such procedure for discriminating between this intelligence and what we subjectively call thought (and, by extension, the entire subjective realm of psyche). On these grounds, it proposes a direct identification of the former with the latter. And finally, it proposes that any systems manifesting this phenotypic property of intelligence, being to that extent objectively indiscriminable, can simply be interchanged; any one of them is a surrogate for any other.

Thus, according to your taste, the Turing Test is either a great step forward or a great step backward. If you believe the former, it constitutes a kind of objectification of thought or psyche itself, bringing it within the realm of hard scientific investigation. If the latter, it constitutes a tissue of equivocations, just as the closely related Church's Thesis turned out to be. Complicating the issue is that the Turing Test adduces "evidence," produced by closely mimicking the very same kinds of cognitive procedures that produce evidence in general, about anything. Thus if something is wrong with the Turing Test, it casts doubt on unthinking reliance on such evidence in general.

Turing framed his initial argument within a very small world, a world of *machines*. This is the world of algorithms, consisting of finite-state hardware executing algorithms embodied in programs or software. Together, the hardware and the programs process inputs, or data, to generate corresponding outputs. In this little world, the *behaviors* so generated, embodied in the relations between inputs and their corresponding outputs, are precisely what constitute *phenotype*.

As described, some of these phenotypes can be effectively discriminated as being intelligent. The actual discriminator may itself consist of such a programmed device (originally, it was supposed to be a human being, but it subverts the whole point of the argument to keep something subjective in this role). The discriminator proceeds by supplying input data ("questions") and evaluating the corresponding outputs, or "answers." In the simplest terms, the classifier is comparing these answers to what its own would be to those same questions. It will thus classify another machine as intelligent or not to the extent that it *simulates the discriminator.* Or, more accurately, simulates it "enough," based on a short list of such questions.

Hence, in this world, the objective phenotypic property of intelligence is identified with "sufficiently simulating" the system that determines the property. Any such discriminator will always, vacuously, classify itself as intelligent—that is, with itself possessing the phenotype in question.

Now this same discrimination procedure, although initially posed in a small world of hardware and software, is not restricted to that world; the test can be applied to other material systems (e.g., to organisms, or to human beings). Indeed, any system to which the test can be applied at all can be said to possess the phenotypic property the test objectifies; i.e., it can be said to be intelligent or unintelligent, depending on the outcome.

In any case, the test provides us with a presumably well-defined class of systems, possessing a certain objective phenotypic character (intelligence), and this class will have at least some machines in it (e.g., the one that actually makes the discrimination). This kind of classification of systems in terms of phenotypic characters or behaviors constitutes the basis for any *taxonomy* and rests entirely on such evidence supplied by discriminators or meters for the characters in question. Indeed, it is generally regarded as *information* about something to specify a taxonomic class to which it belongs (e.g., it is information about an organism to say that it is a mammal; it is information about a particle to say that it is an electron). In the same way, it is information about a machine program to say that it is intelligent.

Indeed, this kind of taxonomic information provides one way of characterizing what a physicist would call the *state* of a system under investigation. A physicist would, from this viewpoint, have to say that such a state always comprises, or expresses, the sum total of all such taxonomic features and is simply a specification of the intersection of every taxonomic class to which it belongs. Unfortunately, that is not the only way of talking about state; there is also a quite different, reductionistic way, and the two are quite different.

The next point to recognize is that we also know, as objectively as anything, the profound limitations and inadequacies of algorithms and machines—i.e., of the universe in which Turing originally framed his test. Especially is this so when one such machine is attempting to assess the behaviors (phenotypes) of another (as in, e.g., the Halting Problem, posed by Turing himself). These considerations lead us to expect that taxonomic classes based on phenotypic properties like intelligence must contain many other things besides machines (which, in themselves, constitute but another taxonomic class). And if so, it brings into serious question whether it is legitimate at all to extrapolate from one *member* of such a taxonomic class to another, simply on the basis of evidence that they share one such class.

Indeed, to the extent that the Turing Test is itself a mimic of traditional experimental, cognitive procedures for classifying phenotypes, it casts serious doubt on those very procedures as a universal, objective way of characterizing material systems in general. It certainly provides no basis for asserting any kind of relation of *surrogacy,* or of *modeling,* between them. In other words, the Turing Test ends up by casting doubt on reliance on any circumscribed set of cognitive modes for this purpose. Accordingly, far from its initial intent to objectify thought by mechanizing it, the test only reinforces and magnifies the need for something other than phenotypic information as a basis for science itself. In the last analysis, what is missing from such pictures is *genotype.*

Above all, it is the possibility of such interchanges between particular nouns and their adjectives that provides the basis for concepts of surrogacy. Thus, for example, as far as turbulence is concerned, oil is surrogate water, and indeed any fluid is a surrogate for any other. Such concepts of surrogacy, or, to call them by their proper name, of modeling, find their ultimate roots in the reality and objectivity of adjectives. It is this reality, which seems perceptible only to an organizing mind, that pure empirics cannot admit. It is indeed primarily this circle of ideas that they condemn as theory.

Reductionism

If there is one thing that has been made clear by the history of biology, brief as it has been, it is that phenotypes entail little or nothing about their associated genotypes. Thus there is nothing in the study of anatomy and physiology, nor for that matter in development or evolution (i.e., in morphogenesis), that even obliquely suggests anything like, for example, DNA. In fact, quite the contrary: the assertion that diverse kinds of systems can still be phenotypically indiscriminable, as the Turing Test claims, clearly argues against any such thing. That is partly why, in biology, genotype and phenotype are so different.

On the other hand, procedures such as the Turing Test are based on a very limited view of what an experiment or observation of a system consists of. They cannot comprehend, for example, experiments involving taking a hammer to a machine's hardware, or putting its software through a shredder. Such experiments radically change everything about a system that the Turing Test assumes preserved; they change the state set, the dynamics, the "transfer function" of the system, in extreme and irreversible ways. Input-output experiments must not be allowed to do any of these things; they must not change the dynamics on which the input-output relation, the transfer function, depends; otherwise, the experiments would be worthless. For this reason, part of the very characterization of a machine asserts the independence of its behaviors (phenotypes) from the experiments used to probe it—namely, that the execution of a program must not change either the hardware or the program being executed. In molecular biology, this taboo is the content of what Crick long ago called the Central Dogma (Judson 1979), that phenotypes must not affect genotype, or that information must not flow from the former to the latter. This dogma is essential in maintaining the mechanical perception of cells that molecular biologists like. But, in fact, this kind of dogma has a far wider currency, which has gone far toward determining the very character of physics itself.

Destructive experiments, which inherently violate all these caveats, serve to create entirely new *systems* from given ones, with new state sets and new dynamics, generally very different from the systems we began with. That is to say, such destructive experiments drastically change the original system's phenotypes, its genotypes, and the relations between them; they destroy everything on which mimesis is based. The *only* relation between the original system and the new one is the fact that they are both composed of the same *material,* the same "stuff" (and even that is not always true). To borrow a term from chemistry, what survives destructive experiments are, at best, relations of *isomerism.*

Thus an initial intact organism is *isomeric,* in this sense, isomeric to a cell population obtained by passing it through a sieve; and each of these cells is isomeric to a spectrum of chemical fractions residing in a hundred different test tubes; and each of these is isomeric to a cloud of electrons, protons, neutrons, or what you will. Each of these isomers is thus treated as a surrogate organism, not because it *behaves* like an organism, but precisely (and only) because it is an isomer of one.

Unlike mimetic strategies, which seek surrogates for organisms via common phenotypic behaviors, reductionism seeks surrogates in a class of ever-bigger isomers, and seeks to express the phenotypes of our original system in terms of those of a big-enough isomer. The ultimate goal is to find a canonical set of "units" such that *every* material system (e.g., an organism) is isomeric to some population of them, and such that every *property*, every phenotype of such a system, can be effectively expressed in terms of those of the units.

The conceptual base of the reductionistic strategy is exceedingly thin, and it has never been of direct conceptual use in dealing with the basic problems of biology (let alone of mind and sentience). On the other hand, it has been exceedingly fruitful in the deployment of procedures for isomerizing things, and in the generation of data about them. The actual pertinence of these activities, however, remains essentially a matter of faith; it is in itself a subjective judgment. (For a detailed critique, see chapter 8.)

The spirit of reductionistic approaches is very much like that of canonical form theorems in mathematics, many of which are variants of the Jordan Canonical Form for matrices. In all these theorems, one needs to know only a limited number of elementary units, and a limited number of ways to put them together; from these, one generates enough "canonical forms" so that everything in the universe of discourse differs from one of the canonical ones only by a subjective thing such as a transformation of coordinates. For instance, a great deal of quantum mechanics rests on the Spectral Theorem, which builds (Hermetian) operators or observables canonically out of projections, and which is in fact a paraphrase of the Jordan Canonical Form to situations that are not finite-dimensional. The interpretation of these projections in terms of observational procedures (in quantum mechanics inherently destructive) is well known, and hence the idea that every property of everything is comprehended within it.

On the other hand, canonical form theorems are useless unless one can specify the transformations that turn the properties of a canonical isomer, in which we are generally not interested, into those of another one in which we are. The very existence of these is primarily a matter of faith, let alone that they are in any sense computable, as Church's Thesis requires. It should be noted that the destructive experiments on which reductionisms depend are inherently irreversible, in any material sense, yet the effect of these presumed transformations is precisely to reverse them, to go backward from a big population of elementary units to a particular material system (e.g., an organism) of which they are an isomer.

Many years ago, I heard a routine of Woody Allen that bears on exactly this point. As he told it, he acquired a Rolls-Royce while in England and wanted to return with it to the States. On the other hand, he didn't want to pay the duty on it. So he hit on the idea of disassembling it, packing the parts into many suitcases, and describing them to the customs inspectors as modern sculpture, not dutiable as art. He was successful, got his many suitcases home, and proceeded to try to reassemble his car. In his first attempt, the parts yielded 200 bicycles. On the second attempt, he got many lawn mowers. And so it went; he never could retrieve the car.

Some Necessary Reassessments

We have seen that there have been two distinct and highly inequivalent strategies utilized for approaching things such as the mind-brain problem, and hence to express subjective things in presumably objective, scientific terms. The first is to replace the system of interest by some behavioral or phenotypic mimic; something that manifests some of its behaviors, but not in general the causal underpinning of those behaviors. Then something like Occam's Razor is invoked to argue that explaining these behaviors in the mimic is adequate for explaining them in the original system. The second strategy is to seek some canonical material isomer, which now behaves completely differently (i.e., it has a different phenotype), but such that every original behavior is a transform of something about that isomer. Indeed, the point of seeking the canonical isomer is to reveal more clearly its context-independent units or constituents; in these units, genotype and phenotype appear to coalesce. The properties of the units themselves are few in number, are themselves known cognitively, and are entirely independent of any larger isomer to which they might belong.

In each case, however, we replace the actual system of interest by some kind of surrogate, and we end up studying that surrogate. It is in this surrogate that objectivity resides, in one form or another, and which is imputed to the original system. But, as we have seen, the basis for this surrogacy is completely different in the two strategies.

These strategies are inequivalent. For example, the first strategy attaches state sets to a system; these state sets are direct sums of its own phenotypic properties, while the second regards state sets as direct products of the units in some canonical isomer. Indeed, in effect, the coalescence of genotype and phenotype amounts to mandating the equivalence of direct sums and direct products, something that is almost never true. Correspondingly, any world in which it is true must be a very special world, a *simple* world (see my *Life Itself*).

Since these two approaches are of such different generality, neither entails the other, nor are they generally intertransformable. Hence, each one by itself is seriously incomplete as a universal scientific strategy.

The inequivalence of these strategies becomes of central importance in biology, where it is literally incarnated in the inequivalence of phenotype and genotype. We might say that this inequivalence is in itself a phenotypic property manifested by the class of material systems we call organisms; it constitutes as good a diagnostic of the living state as there is. From the beginning, I have started from this separation between genotype and phenotype; it is the very point of departure of the relational models I have called (M, R)-systems (see chapter 17). It also indicates why organisms resist reductionism so mightily, since in a sense the point of reduction is to dispense with that very property.

For such reasons, I argue that life poses the most serious kinds of challenges to physics itself, orders of magnitude more serious than the ones electrodynamic phenomena posed for Newtonian mechanics, or that spectra and chemical bonding posed for nineteenth-century classical physics. More specifically, the expectation that phenomena of life or mind could be assimilated directly into physics as merely a minor technical bubble, of no conceptual significance, was mistaken. In a certain sense, the problem is that contemporary physics has never, of its very nature, had room for a concept of genotype. It can maintain its present form only by staying close to material systems in which genotype can be ignored; in biology we meet material systems where it cannot be ignored any more. But this requires a radical conceptual restructuring of physics from the ground up.

If one wishes to call the incorporation of genotypic concepts into physics a kind of *complementarity,* à la Bohr, there is no objection. But this does not help physics itself actually do it.

At heart, the concept of genotype addresses the causal basis of phenomena, or phenotypes; phenotypes are *effects*. What I am suggesting is that these causal bases are themselves not *just* other phenomena, and hence, in particular, that phenotypic things such as life and mind are not *just* another way of perceiving, or apperceiving, the small repertoire of cognitively visible, context-independent properties of a few reductionistic units. Nor can the phenotypes be separated from their causal bases, as is tacit in mimetic approaches such as the Turing Test.

These problems occupy center stage in biology. But of course they have antecedents and predecessors in contemporary physics. They are there treated as technical anomalies, raising no issues of principle, just as was, say, the anomalous specific heat of the diamond, or the behavior of the perihelion of Mercury. Only in hindsight could it be recognized that such anomalies were foretokens of revolutionary upheavals, rather than technical difficulties that could be resolved with just more cleverness in applying received principles. So, for example, most physicists dismiss the measurement problem in quantum mechanics as something of no importance, even though it is, in a sense, an embryonic form of the mind-brain problem.

The problem here is that the very acquisition of data, the very cognition of phenomena (phenotype) in a material system, requires one to consider a *larger* system ("system + observer") and not to consider smaller ones, as reductionism (or context-independence, or objectivity) requires. This in turn creates a chicken-egg situation, an *impredicativity;* specifically, one must know the larger system to characterize the smaller, but one cannot know the larger until the smaller is characterized. Contemporary physics does not like impredicativities in nature, but of course it is deeply concerned with preserving the objectivity of the cognitive processes (e.g., measurements) on which it bases itself. Pais (1982), in his assessment of this problem, declares it only a game played by mathematicians and epistemologists, devoid of impact on *practical* physics. But then, practical people tried to refute the Zeno paradoxes by getting up and walking around the room.

Nevertheless, the excluded process of finding out about a given system by putting it into a *larger* system (i.e., by creating a definite context for it), in the teeth of the impredicativities this generates, is perhaps the only exit from the boxes that reductionism alone, or mimesis alone, creates for us. Biology already forces us out of these boxes, for life and mind are not to be found within them.

The pervasive fear of impredicativities, and hence the attempts to encompass all of objective reality within purely predicative bounds, is quite needless. To be sure, one must be careful with them; they can lead to paradoxes if one is not. But biology clearly shows that one cannot simply dispense with them, in physics any more than in mathematics.

Indeed, to capture something like the genotype-phenotype dualism in purely predicative (i.e., algorithmic or computable) terms, or, what turns out to be the same thing, to identify direct sums with direct products, generally requires an infinite number of these algorithms. No canonical forms of the type described in the preceding section exist for systems of this type, which I have termed *complex.* In effect, contemporary physics is based on the belief that *no complex systems exist in nature.* This is but another form of Church's Thesis, and I argue that it is false. On the other hand, if it is false, then there are plenty of material systems (organisms) whose phenotypes can be neither regarded as a mere recognition of a limited set of properties of reductionistic units, nor exhausted by a limited number of mechanical mimics. Either of these commands a fundamental enlargement of our physical perceptions of material nature itself.

Some Remarks on Inertia and Gravitation

The fact that effects do not in general entail their own causes (i.e., that phenotypes do not entail their own genotypes) is central to the entire preceding discussion. If this is so, then merely cognizing the phenomena that populate our objective world of events and behaviors is not enough; it is necessary, but almost never sufficient. One way of expressing the difficulties of both reductionism and mimesis is to note the underlying belief that enough necessary conditions comprise a sufficient one.

Thus neither of the following is true, that enough cognitions of brain will suffice to make a mind, or that enough behaviors of mind will suffice to make a brain. The belief that there is always a threshold, embodied in the word *enough,* that will make such assertions true is in its turn a consequence of reliance on repetition of a limited repertoire of rote processes, and on the belief that enough such repetitions will take us to any threshold. Repetition of rote processes is of course the essence of algorithm. And as we have seen, algorithms embody a world without impredicativity, a world in which genotype and phenotype coincide.

In such a world, analysis and synthesis are simple inverse processes. There is no difference in describing a system in terms of how its own behavior is determined by its subsystems, or what that system does to larger systems to which it might belong, and if there is no difference, we might as well concentrate on the former. This is part of the predicative view, that we must look only to subsystems to understand something, and never try to understand it by looking to larger systems, or specific *contexts.* As we have seen, this not only creates impredicativities but seems to deny objectivity or context-independence itself.

Therefore it is instructive to hark back to the foundations of classical Newtonian mechanics, a kind of quintessence of objective science, and in particular at how it treats (1) the *response* of a particle to an (external) force impressed upon it, and (2) how that same particle can *exert* a force upon other particles.

The problem (1) is addressed, in one fashion or another, by Newton's three Laws of Motion. The problem (2), on the other hand, requires a completely separate and independent Law of Universal Gravitation. The latter has nothing to do with the former; it cannot be entailed from Laws of Motion, nor they from it.

Now, how a mechanical particle responds to a force is tied to its *inertia* and is embodied in a specific parameter called *inertial mass.* On the other hand, how that same particle *generates* a force is embodied in a conceptually quite different parameter, called *gravitational mass.* The former is phenotypic in nature; the latter is genotypic.

It turned out that, although these parameters are very differently defined, on the basis of entirely separate laws, the *values* of these parameters were numerically the same. This strange fact was later seized upon by Einstein: he regarded it not as a happy coincidence but rather as an expression of something very deep about physics itself. He ultimately expressed it in terms of a Principle of Equivalence between inertia and gravitation, asserting the nonexistence of any experiment, done entirely within a mechanical system, that could discriminate between the one and the other; this ended up as a cornerstone of General Relativity. In our terms, it asserts the fundamental identity of genotype (gravitation) and phenotype (inertia) in mechanics.

But even here, this stops being true once we try to interpret, or *realize,* Newtonian impressed forces as arising from the gravitation of other particles; that is, as soon as we put our particle into a bigger system of particles. Roughly, when we do this, we are allowing that impressed force to be determined (at least in part) by the particle itself. The particle is now (phenotypically) behaving in a "field of force" that it participates in generating. Although a classical particle is forbidden from (gravitationally) pushing itself when it is alone, adding other particles allows it to push things that can push it. The resultant impredicativities, which in fact plague all field theories of particle-based forces, arise as soon as we try to identify inertia (phenotype) with gravitation (genotype) independent of any larger context.

This is why, for instance, one cannot solve a three-body problem reductionistically, by solving two-body and one-body problems.

The impredicativities I have described have bothered physicists for centuries, because even here, they involve an intertwining of genotypic and phenotypic considerations, whose absolute separation (or absolute identification) is at the heart of the entire picture. They deeply disturbed Ernst Mach, for example, and led him to propose a view of mechanics quite different from, and incompatible with, the Newtonian, based entirely on phenotypes. In practice, most physicists ignore such problems, through the employment of mean-field approximations that serve to again isolate the inertia of a particle from the effects of its own gravitation.

The capacity for self-forcing, the effects of a particle's gravitational properties upon its own behaviors when there are other particles around, can therefore not easily be expressed by differential rate equations, or dynamical systems, at all. They are, rather, expressed by networks of interlocking differential forms that describe direct and indirect effects on inertia of gravitation, what I have called *activation-inhibition networks* and *agonist-antagonist networks,* for example (Rosen 1983). These networks of differential forms collapse into a finite system of ordinary rate equations, or equations of motion, when all these differential forms are exact, but not otherwise. It turns out that the nonexactness of some or all of these differential forms is another way of describing *complexity,* as I have used the word.

Thus even in classical mechanical physics, any presumed objective identification of inertia and gravitation, or of genotype with phenotype, already raises deep questions bearing on the adequacy of both mimetic and reductionistic strategies for coping with material reality. These problems, and many others like them, become critical in using those strategies to cope with life and with mind.

What has all this to do with the mind-brain problem, our initial point of departure? I suggest that contemporary physics, the science of material reality and the touchstone of objectivity rooted entirely in cognition, rests on a host of presumptions that in themselves not only are noncognitive in nature but need not hold (and, I argue, do not hold) in systems that live and think. There is nothing "unphysical" in acknowledging these facts, nor does it mean an abandonment of the laws on which it is presently based; it means only that, in addressing life and mind as material phenomena, which they are, we cannot expect to find only specializations of these laws, but rather generalizations of them.

I will try to illuminate these last assertions by looking at cognate situations in mathematics—in particular, at what I have called "real" number theory and its various formalist syntactic simulacra (which can be called *artificial* number theories, sometimes *constructive* number theories). It is a primary upshot of Gödel's Theorem that the real theory is always much bigger than any of the artificial, predicative, algorithmic, machine-based ones. Nevertheless, the real theory does not violate any of the laws on which the artificial ones are based. But it does not specialize these laws; it rather enormously *transcends* them. On the other hand, there is no purely phenotypic *evidence* which can effectively discriminate the real from the artificial. Certainly, this discrimination cannot be made from entirely within such an artificial arithmetic alone. To do so requires a radical enlargement of the cognitive base of the system itself, and a corresponding enlargement of their genotypic foundations.

A Brief Summation

We have proceeded from the fact that nobody doubts, or even *could* doubt, his own life and his own sentience. Nor does anyone doubt that these things are inseparable from, and indeed are properties of, a certain material system, which he calls his body. That makes them, in biological parlance, *phenotype,* even though the knowledge of them is noncognitive. Nevertheless, they should be amenable to the general procedures used by biologists to study phenotype, and, more generally, to the procedures used by physicists to study material systems. But none of these procedures, individually or collectively, has shed much light on either the basis of life or of mind.

So something is wrong. There are two possibilities: either the procedures are universally sound, and we have merely been remiss in carrying them out properly, or the procedures are limited or flawed. In the latter case, no amount of industry or diligence based entirely on these procedures will suffice. I argue for the latter possibility.

Part of the problem here is that evidence alone, short of actually solving these problems, does not suffice; such evidence supplies at best necessary conditions, and not sufficient ones. Nevertheless, like everyone else, I have throughout this discussion adduced such evidence.

An important part of this evidence is historical in nature. The fact is that neither reductionistic approaches to our problems, anchored in (theoretical) physics, nor mimetic approaches have shed any real light on the problems themselves. As to the former, whenever theoretical physics ventures into the biological realm, either it has nothing to say, or what it says has so far been wrong. Historically, whenever this happens, it has bespoken a Foundation Crisis of the first magnitude in physics. It has already happened many times. Why should it be any different this time?

As we have seen, a *reduction* of biology to contemporary physics throws the problem onto finding some canonical isomer of an organism, and expressing the latter's phenotypes entirely in terms of independently known, cognitive, observable properties of the units constituting that canonical isomer. The fact that every property of every other isomer must also be so expressible simply sweeps the problem under the rug of "initial conditions." In my view, this simply begs the entire question—a concession of abject scientific impotence.

Nevertheless, the difficulty of finding the right needle of initial conditions in the haystack of all possible ones provides the basis for the common view that organisms constitute a rare and negligible specialization of physical universals we already independently know. To suggest otherwise is to expose oneself to charges of vitalism and worse.

To the contrary, organisms vastly generalize the cognitive bases on which contemporary physics rests, in the same way that real arithmetic generalizes any purely syntactic artificial formalization. In that sense, one does learn deep lessons about *physics* by grappling with phenomena of life and of mind.

This discussion is framed in terms of the distinction in organisms between phenotype and genotype, whereas physical approaches are based on a mandated coincidence or coalescence of the two. Their distinction in organisms is reflected in the fact that no amount of phenotypic evidence entails anything about the underlying genome.

An essential feature of this discussion is the necessity of learning about a material system, not only by taking it apart into subsystems, but also by putting it into larger systems with which it can interact. This was the substance of the preceding discussion of inertia and gravitation. In biology, these two ways of viewing a given material system are already radically different, and highly inequivalent. But as we have noted, the invocation of specific larger contexts creates *impredicativities* that contemporary physics does not like to admit; that is precisely why there is a measurement problem in quantum theory, and a life-organism problem in biology, and a mind-brain problem. Such problems ultimately force us into some deeply ingrained habits of thought, which it is entirely possible are bad habits.

Schrödinger (1944; see chapter 1) argued forcefully that biology mandated "new physics." Both physicists and, today, molecular biologists are (for quite different reasons) vehement in their denial of any such thing. But today, fifty-five years later, when conventional approaches have failed to move us an inch closer to problems of life and mind, such exhortations may begin to seem a bit more plausible.

Addendum: Some Concluding Remarks

In the preceding discussion, we have taken an exclusively *synchronic* view of the mind-brain problem—a partition of the present world, now, today, into an objective part and a subjective one. The former is the province of *physics* in the broadest sense, a world of phenomena or phenotype, in which subjectivities are allowed no causal role. The mind-brain problem, paradoxically enough, seeks to pull mind and awareness, the basis of subjectivity, back into this external objective world.

One powerful motivation for trying to do this (quite apart from the subjective lust for unification) is found not in synchrony, but rather in *diachrony*—in history or, as the biologist would say, in phylogeny. It is possible to imagine a time before consciousness, a time before life. At such a time, there was *only* an objective, phenomenal world, and hence no separation or partition between objective and subjective. The partition itself, according to this view, had to *emerge* diachronically from the preexisting, abiotic, amental objectivities, through a process of evolution. Hence, one important aspect of the mind-brain problem lies in running this diachronic process, this history or evolution from abiotic to biotic, backward. Specifically, to try to pull the subjective into the objective by pulling the present into the sufficiently remote past.

Such a diachronic approach is different in spirit from either mimesis or reductionism, neither of which is diachronic in this sense.

To consider this approach in detail is not possible here. I will make only a few remarks about it, in terms of the analogy between *objective* in the causal universe and *formalizable* in the mathematical one. This latter, of course, does not have a diachronic aspect in the same sense, but it does have a historical one. The very concept of formalization was an exceedingly late one, first expressed by Hilbert, but with roots going back to Euclid and before. Hilbert, of course, thought initially that everything in mathematics was formalizable, but Gödel very quickly showed that this was not so—indeed, that formalizability was rare in mathematics. Diachronically, in fact, mathematics ran quite the other way: it did not emerge from objective, formalizable beginnings, to successively more complex, nonformalizable elaborations. There was no such process of evolution in the history of mathematical thought. If anything, this evolutionary process ran quite the other way. In particular, the formalizable, or simple, systems were not intermediate evolutionary stages, or building blocks, in the elaboration of more complex, nonformalizable mathematical systems. Nor, as Gödel showed, are they even useful subsystems in studying their synchronic properties, their physiology, if you will.

My suggestion is to take this mathematical metaphor seriously in the material world as well—namely, that diachronic evolutionary processes do not necessarily go from simple to complex in matter, any more than they do in mathematics.

CHAPTER 7

ତ

On Psychomimesis

In chapter 5, I discussed the mind-brain problem (and the closely related life-organism problem) as attempts to express subjectivities (mind and life) in presumably scientific, objective terms (brain and organism). I drew parallels with the notorious measurement problem in quantum mechanics on the one hand, and attempts to root the objectivity of mathematical truth entirely in syntax on the other. I roughly concluded this: Any attempt to reconcile the contradictions between subjective and objective on entirely objective grounds must fail as long as we persist in identifying objectivity with machines, and with syntax, alone. Indeed, not only can the mind-brain problem not be solved in such a world, it cannot even arise.

In *Life Itself* I described the kinds of larger worlds, the worlds of complex systems, in which such questions can meaningfully be formulated, addressed, and solved. I shall not repeat any of those arguments here. Rather, I shall concentrate on one ancient strategy for trying to incorporate the subjective in the objective—the strategy of *mimesis* (see chapter 6). The idea is that a material system manifesting "enough" properties of organism *is* an organism; a system manifesting "enough" properties of mind *has* a mind. These endeavors have always been closely linked. As the former has been termed biomimesis, it is reasonable to call the latter *psychomimesis*.

The relation between mimetic approaches (manifested today in such things as artificial intelligence and artificial life) and the quite different "scientific" approaches (embodied, e.g., in neurophysiology and molecular biology) is itself one of mimesis. Mimesis involves synthesis—the pasting together of behaviors or properties until one has enough to claim identification with life or mind. Scientific approaches tend to proceed analytically, starting with a particular system that lives or thinks, and then taking apart into a spectrum of fragments or subsystems, seeking to express the original system's life and mind in terms of those fragments, which have neither. In a deep sense, mimesis violates the reductionisms on which scientific analyses depend, because it inherently abstracts the subjectivities away from the particular systems that manifest them, an abstraction that is analytically inadmissible. Conversely, those who practice mimetics claim their own kind of objectivity and tend to look on the analytic particularities of the biologist as quite irrelevant.

The two enterprises mimic each other because they pursue the same goals or ends through entirely different means. This, as we shall see, is the essence of mimesis itself. They are not reconcilable because, among other things, analysis and synthesis are not in general simple inverse processes: a synthetic process is not an analysis run backward or conversely (Rosen 1988b).

Nevertheless, science and mimesis have continually intertwined in subtle but profound ways. Indeed, the "mechanism" espoused by scientific biologists today, and whose negation they mistakenly identify with vitalism, derives largely from the Cartesian machine metaphor (see chapter 20, and *Life Itself*). This in turn rests on Descartes's mimetic assertion that, because machines can sometimes manifest lifelike behaviors, organisms *are* machines.

As I shall employ the term in what follows, *mimesis* involves classification of things on the basis of a perceived sharing of properties, or behaviors, or attributes, independent of any shared causal basis for these attributes. Science, on the other hand, is concerned with the causal underpinnings of those attributes on which mimesis depends; it classifies systems according to these underpinnings and cannot rest content with the seductive superficialities of mimesis.

Thus science says that the Monarch butterfly and the Viceroy butterfly, which closely resemble each other in size, pattern, and coloration, are in fact not very closely related at all. The mimicry, or mimesis, exhibited between them does not extend to any deep relation between any of their other properties or behaviors.

On the other hand, there is a very strong subjective tendency to extrapolate from a mere mimesis to the sharing of other properties. In the butterfly example, for instance, even a potential predator is credited with just such an extrapolation, in effect reasoning that a butterfly looking enough like a Monarch *is* a Monarch, and hence to be avoided (Monarchs, unlike Viceroys, taste bad). Such inductions are, in fact, examples of what we have called *anticipatory behaviors* (Rosen 1985a), but this is not my present concern.

This argument from a mimesis to an identity is a version of the Turing Test, designed by Turing (1950) to provide an answer to the question, Can a machine think? and thereby to provide an operational characterization of thought itself. It is precisely this kind of extrapolation that allows mimesis to encroach on science and even on mathematics, as we shall see. In fact, the real question, approached mainly in the context of machines, brains, and mind, is the extent to which mimesis is in itself science or is only a mimic of science.

On Mimesis

The Greek word *mimesis* is generally translated as *imitation*. Consider, however, the opinion of W. K. C. Guthrie (1962) quoted in chapter 4, and the following passage from the same source:

> *Mimesis* meant acting as much as imitation, *mimetes* was often and *mimos* always an actor. The relation between an actor and his part is not exactly imitation. He gets inside it, or rather, in the Greek view, it gets inside him, and shows forth through his words and gestures. There is more to it than that. Drama began as, and remained, a religious ritual. . . . In the earliest and simplest dramatic representations men impersonated gods or spirits, and what they supposed to be happening can be best illustrated from contemporary ecstatic worship like that of Dionysus. The leader of his *thiasos,* the band of god-intoxicated worshipers, impersonated, or imitated, the god. So we might put it, but to him and his fellow-worshipers what happened was that the god himself entered into him, took possession and acted through him. . . . In myth the god was attended by a band of *daimones,* and in performing the ritual the worshipers not only acted the parts of, but for the moment *were,* the god himself and his divine attendants—Bacchoi, Kuretes, Korybantes, or whatever the name might be. (p. 230–31)

This is a very ornamental, mystical language, but it is essentially the Turing Test, that things that *act* enough alike are to be identified. Men who *act* enough like gods *are* gods—without regard for causality or anything else. It was precisely this neglect of causality, this emphasis entirely on the *acting,* that underlay Aristotle's disdain for both Pythagoreanism and for Platonic Idealism.

Seen in this light, mimesis really goes back long before the Greeks, to the most ancient roots in sympathetic magic. It may seem extravagant to regard something like the Turing Test as a modern embodiment of the archaic occult concept of "Sympathies," but there it is. (As a matter of fact, there is even a similar magical element involved in the reductionistic, scientific belief that, because a part has been isolated from a whole, therefore the properties of the part are transmitted to the properties of the whole. Hence that enough properties of enough such parts in effect *constitute* the whole.)

In such terms, the general-purpose digital computer, or the Turing Machine, is regarded as the universal actor, which can behave like anything if provided the right script (program or software) . . . and, therefore, it *is* everything.

On the other hand, we can contrast these notions of mimesis with another, traditionally considered interchangeable with it. Etymologically, the word *simulation* derives from the Latin *simulationis,* which connotes a feigning or pretense, a sham or counterfeit, based on irrelevant semblance. In this light, it is a very different thing to say that the Viceroy mimics the Monarch, or that it simulates the Monarch. We merely note here that inductive discriminations, such as the Turing Test, cannot tell the difference between the one and the other, however different they may be conceptually.

Biomimesis and Psychomimesis

Life and mind have been linked since earliest times. The modern word *animal,* which pertains to life and the living state, derives from a Latin word meaning both breath and soul. The corresponding Greek word is *psyche.*

On the other hand, *organism* derives from the Greek *organon,* which means tool, implement, or engine.

The jarring apposition of these words reflects the apposition between mimesis and science in trying to come to terms with concepts such as life and mind. As I have indicated, science tends to proceed analytically and reductionistically, starting from actual specimens, and counting on the origin of an analytical fragment from the intact specimen to ensure the relevance of studying the former to learning about the latter. The mimetic approach, on the other hand, seeks to incorporate more and more of the behaviors of such a specimen into another system, not derived from the specimen itself, hoping to cross some threshold ("enough") where the behaviors of the two become indiscriminable according to some decision criterion, and hence the two systems can be identified.

D'Arcy Thompson, in his famous book (1917), gives some of the flavor of biomimesis (more accurately, cytomimesis) as it was pursued in the last century. The emphasis was on producing artificial cells, working with oil droplets in various ionic baths. The idea was to concentrate on mimicking more and more cellular behaviors such as motility, irritability, and the ability to follow physical or chemical gradients (tropisms). Some of these artificial systems were actually quite startling. Such approaches still appear as scenarios in the literature on the origin of life.

Long before such chemical approaches, mimesis was pursued mainly in mechanical contexts, in terms of clockworks. When galvanism was discovered, it quickly became the fashion, and in the course of time a whole family of different "neuromimes" were characterized (Harmon and Lewis 1966). Mimesis was from the beginning a frank object of the field of cybernetics (Wiener 1961), as was manifested early in, for example, the light-seeking "tortoise" of William Grey Walter. Norbert Wiener (1961) provided the ultimate in mimetics with his idea of reproducing every behavior of an organism by, in effect, setting coefficients in a function generator. He thus proposed a biological version of the Ptolemaic Epicycles as his solution to the problems of life and of mind—another version of the perfect actor, whose scripts were now coefficient arrays or parameter vectors.

Mimesis was also involved in trying to deal with questions pertaining to reductionist parts, as in trying to specify what made enzymes catalysts. There was, for example, a fairly extensive literature on enzyme models—inorganic catalysts that catalyzed similar reactions. Thus colloidal metal suspensions were invoked to study catalases and peroxidases, because they acted alike.

These few illustrations should suffice to give the flavor or these approaches; many, many others could be adduced. The most recent efforts in these areas, collected under the rubric of artificial life (Langton 1989; Langton, Taylor, and Farmer 1992), are entirely in these mimetic traditions, expressed mainly now in terms of machines and programs—hardware and software. It is curious that many of those presently pursuing such ideas seem quite unaware of the ancient tradition they perpetuate.

Modern Psychomimesis

For a long time, biologists were convinced there could be no "theory" of the brain, or of the mind. The mind's mysteries—cognition, volition, consciousness, and whatever else constitutes psyche—so immediately apprehensible to each of us, were forever lost in the unfathomable complications of the brain. As evidence, people would trot out the microscopy of Santiago Ramón y Cajal (1933), pointing out that the complications manifested in those infinitesimal slivers had to be multiplied by at least ten or twelve orders of magnitude.

Compared to such problems, the available resources seemed puny indeed. There was neuroanatomy. There was neurophysiology, which concentrated on the generation and propagation of action potentials. On another and apparently quite separate front, there was psychophysics, dealing with behaviors mediated by brain: learning, discrimination, memory. There was the study of "simpler" brains in other species and other phyla. All these are still with us, some of them exquisitely refined, but still of the same character; they remain the backbone of a "scientific" approach, an analytic, reductionistic approach to what brains are and what brains do.

A radical transformation took place in the early 1930s, when Nicolas Rashevsky conceived the idea of neural networks. He was then interested in phenomena of excitation and propagation in peripheral nerve fibers, which is where the data were. Like everyone else at that time, he had to proceed entirely phenomenologically from those data. His approach was to atomize peripheral nerves into what he called excitable elements. The data suggested what the input-output behavior of such an atomic element would have to be, or what its "transfer function" was. He then converted this input-output description into another one, in terms of internal state variables—the first example of this procedure that I know of. Since only two state variables were formally required, one for excitation and one for inhibition, he called his excitable atoms two-factor elements.

At this point, conventional neurophysiologists lost interest, since Rashevsky could not tell them what his factors were, or how they could be independently measured. He nevertheless proceeded to turn these ideas into an approach to peripheral nerve propagation, by the simple expedient of stringing his excitable elements in series.

The crucial idea was the observation that excitable elements could be wired together in more general ways than in simple series circuits: they could be arrayed in networks. The elements in such networks were clearly neuronlike; the appearance of these networks were clearly brainlike. The crucial observation, however, was Rashevsky's exhibition of quite small networks of this kind that could actually *behave* in brainlike ways, networks that could legitimately be said to discriminate, to learn, and to remember.

Rashevsky was well aware that he was proposing a mimesis, and not a "scientific" theory, of brain. He thought of it as a kind of existence proof, to counter the arguments that brain was unknowable, and psyche forever lost in that unknowability. However, it was technically difficult to study behaviors of any but the simplest networks; his two-factor elements manifested hard nonlinearities, analytic procedures were out of the question, and he would not have used computers even if they were available (see Rashevsky 1960).

In 1943, about a decade after Rashevsky developed these ideas, his student Walter Pitts, together with Warren McCulloch, paraphrased the two-factor nets in a discrete, Boolean context, and thereby replaced analysis by algebra (McCulloch and Pitts 1943). The neural net thus became a switching net, a technological thing. This was an essential, if largely unacknowledged, ingredient in what crystallized into "artificial intelligence" a decade or two thereafter.

The idea of the neural net quickly became entangled with others, some from technology, some from the foundations of mathematics itself, into an area loosely called automata theory. Two of the more prominent people involved in that entanglement are Alan Turing (1950) and John von Neumann. They were both intrigued by the dual role of mathematics as a product of mind and as a describer of brain. Especially were they drawn, in their separate ways, to notions of algorithm and its embodiment in software and programs.

In the beginning, the neural net was a plausible mimic of brain, because it did not depend entirely on a comparison of the behaviors of each; it rested rather on anatomy. Indeed, as originally conceived, the behavioral mimicry arose from the underlying mimicry between biological neurons and switchlike elements, and on a continuity assumption, or robustness hypothesis, that populations of comparable elements arrayed comparably would behave in comparable ways. This kind of plausibility has been entirely lost in the progression from neural net through finite automaton through Turing Machine, in which comparability devolves entirely on behaviors themselves, rather than on the way the behaviors are generated. In mimetic terms, we now have actors (e.g., Turing machines) imitating actors (automata) imitating other actors (neural nets) imitating brains. What looks at each step like a gain in generality (i.e., more capable actors) progressively severs every link of plausibility and throws the entire burden on the actions alone.

The effect of something like the Turing Test is to assure us that this procedure of comparing actions alone is all right, and indeed more than all right.

Mimesis and Side Effects

As we have seen, mimetic strategies rest on the supposition that, if a system and a mimic have enough behaviors in common, they must have other (indeed, all other) properties in common as well—properties other than those on which the original mimicry is based. This extrapolation from some to all, made entirely on behavioral grounds and devoid of any shred of causality, is where the magic is.

As a prominent example of this kind of approach, we have considered the Turing Test, which says that if *we* cannot discriminate behaviorally between another human being and a properly programmed machine, then the two must be *objectively* identified. In short, we must *impute,* to anything that behaviorally mimics a mind or consciousness, its own mind and its own consciousness; we must impute to it a subjective, inner world just like ours.

Turing's Test thus asserts that mind, or awareness, or consciousness, is coming along as a kind of *side effect* of a sufficiently elaborate program to a digital device. As I described it in *Anticipatory Systems,* a side effect is an unplanned, unforeseeable consequence of controls imposed on a system to accomplish some other purpose. I treated such side effects as discriminators, rooted in system behavior itself, between what we are actually doing and what we *think* we are doing (on the basis of a model or predictor). As such, the manifestation of a side effect means bad or inadequate models.

On the other hand, the mimic in this case is a machine, or a mechanism—a purely syntactic object, which executes, and can be completely described by, algorithms. It is the kind of system I have called (in *Life Itself*) simple. A basic property of such simple systems is that they possess largest models, which are themselves necessarily simulable. When we use these largest models, there can be no side effects. Indeed, every aspect of system behavior, and the generation of behavior, is explicitly embodied in the finitely generated syntax of that largest model. In principle, this is available for the machine we are using to mimic some, or enough, of the behaviors of mind. It is not available for the system being mimicked—that is, for us. But just as Turing's Test imputes mind to the mimic, it imputes simplicity to that which is mimicked.

However, side effects in general indicate, as we have seen, more system models than we have available. Turing's Test can be interpreted as exhibiting mind as a side effect of enough behavioral programming to fool some set of discriminators. But that programmed mimic, being simple, has a largest syntactic model; there cannot be a larger one, as required by mind as a side effect. Thus either Turing's Test is false, or the putative mimic is not simple—it is *complex.*

In either case, there are never enough purely behavioral discriminators that *entail* identity. We must go to causality for that kind of identification—i.e., into how behaviors are generated. One immediate conclusion is that we cannot generally argue backward from behaviors alone to their causal bases (i.e., analysis and synthesis are not simply inverse operations [see chapters 1 and 6]). Behavioral mimesis is at best only evidentiary; it has been argued that inductions from such evidence, like the Turing Test, have no basis, since at least the time of Hume.

Ideas about the inference of identity from behavioral evidence show up already in physics, as manifested for instance in the Gibbs paradox. I argued a long time ago that determining whether two gases were "identical" or not, on the basis of discriminators, involved solving a word problem (i.e., it was algorithmically unsolvable), and hence in particular that entropic computations about them were unreliable. Physicists (e.g., Schrödinger) tend to simply beg the question by mandating that elementary particles *are* identical.

Generally, we can try to operationally discriminate between a system and a mimic by having one of them interact with a *larger* system; we can observe the latter's behavior and then simply replace the original system by the mimic. If *we* can discriminate a difference between the two situations, then the mimicry fails; that larger system can already tell the difference between the two.

In the case of the Turing Test, such a discrimination is provided by Gödel's Theorem. This allows us to discriminate between a formalization of Number Theory (i.e., an artificial arithmetic generated on the basis of pure syntax) and "real" number theory (a frankly subjective product of human thought). Gödel's Theorem may be regarded as exhibiting the *complexity* of Number Theory, manifested in impredicativities and in semantic features (external and internal referents) that are by definition excluded from any syntactic mimic. And if we can thus discriminate between artificial and real mathematics, we thereby discriminate between an artificial and a real mathematician.

Conclusions

Mimesis is based on the idea that if two systems act enough alike, they can be identified. We have explored this idea mainly through the Turing Test, asserting that a properly programmed machine, operating via syntax alone, that behaves enough like a thinking human being *is* thinking. By extension, then, the argument is that every subjective property of mind or sentience is in fact present in a sufficiently programmed syntactic device.

As far as actually understanding life and mind is concerned, I claim that mimesis is more akin to sympathetic magic, and as such it has primarily a recreational value. It is only when behavioral mimicries can be *extended* to the causal underpinnings of behaviors that these mimicries are of relevance. Specifically, in exploring what machines can do as a probe of mind, I argue that no such extensions exist. In other words, that fields such as artificial intelligence deal no more with intelligence than, say, symbol manipulation deals with literature or poetry.

Closed, syntactic worlds are, by their very nature, extremely poor in entailment relations, and hence in their capacity to represent or model causal relationships. In effect, inductions such as the Turing Test are an attempt to compensate for this causal impoverishment without actually enriching those worlds.

On the other hand, it is also true that a complete identification of causality with syntax and algorithms, in the name of objectivity, is itself a very human thing to do. The idea that enough iterations of rote operations will carry one across any threshold is an attractive one, and it was forcefully argued (e.g., by von Neumann) under the rubric of complexity. But it will not, for instance, carry us from finite to infinite, nor yet back again. It is indeed a very *subjective* thing to do, especially when more attractive alternatives are available.

∾

CHAPTER 8

The Mind-Brain Problem and the Physics
of Reductionism

Discussions of the mind-brain problem are inseparable from the capabilities and limitations of reductionism. All problems of this type possess the same basic form: Given a material system x, such as an organism or a brain, we want to answer a question of the form, Is x alive? or Is x sentient? or, in general, Does x manifest or realize or instantiate a property P? That is, we want to treat P as a predicate, or adjective, of a given referent x, just as we do, for example, specific gravity, or reactivity, or the shape of an orbit. Stated otherwise, we want to determine the truth or falsity of some (synthetic) proposition $P(x)$ about x. A rather different (though closely related) question may be patterned after the title of Erwin Schrödinger's 1944 essay, What is P? This last question merely regards x as an instance of P, rather than regarding P as an adjective of x.

The totality of true propositions $\{P(x)\}$ about x constitute one way of talking about the *states* of x. We shall come to another, quite different way in a moment. It is one of the basic tenets of reductionism that the different ways of characterizing the states of any x must coincide.

Reductionism, as a general strategy for answering such questions about specific material systems x, actually asserts several things. First, and most weakly, it mandates that any objective property $P(x)$ of x be expressible as a conjunction of a finite number of other properties $P_i(x)$, each one necessary, and all together sufficient, for $P(x)$ itself to be true. It goes without saying that the truth or falsity of each of these constituent properties $P_i(x)$ must be determined on purely cognitive grounds—ultimately, through sensory impressions or observations, brokered or transduced through appropriate measuring instruments. That is, there is a finite list of such conditions P_i, an *algorithm,* that can be programmed into a finite-state "machine," according to which we (or it) can decide whether a given system x is "alive" or "conscious," or, in general, whether $P(x)$ is true or not.

Thus reductionism asserts a kind of fractionation of an arbitrary predicate or property of a material system such as x, as a conjunction of essentially independent subproperties P_i, so that we can write the following:

$$P(x) = \bigvee_{i=1}^{N} P_i(x), \qquad (8.1)$$

where the truth or falsity of the propositions $P_i(x)$ can be determined on purely cognitive grounds. Moreover, as asserted above, the truth of the constituent summands on the right-hand side of equation 8.1 are each individually necessary, and are all together sufficient, for $P(x)$ to be true. Hence, in particular, the property P is redundant and need not separately be included as part of the "state description" of x itself; it can be recaptured from the P_i by purely syntactic means. That is indeed the essence of equation 8.1. Stated another way, the truth of $P(x)$ does not constitute new *information* about x.

Despite a great deal of effort over the years, no such list, fractionation, or algorithm has ever been forthcoming as an answer to the mind-brain problem, to the mind-organism problem, or to other problems of a similar character. There is no shortage of proposals along these lines, but they have proved to constitute neither necessary nor sufficient conditions that a material system be alive or sentient, or, indeed, that a truly general proposition $P(x)$ about x be true.

There are two ways to interpret the profound absence of such lists, which reflect the failure to establish P as a conjunction of the form of equation 8.1. The easiest interpretation is that equation 8.1 is sound but that we simply have not found how to fractionate P in this form *yet.* The other, more radical one is that the entire strategy is faulty— i.e., that there are properties P that cannot be fractionated in the form of equation 8.1 at all. In this case, a property such as P does indeed provide information about x and must be an essential part of its state description. In this case, there is no list or cognitively based algorithm for deciding the truth of $P(x)$ given x. This possibility already raises the specter of *noncomputability;* it says something drastic about material systems and their properties. That is, it says something drastic about our contemporary views of physics itself (where by *physics* I mean the science of material systems in all of their manifestations).

Schrödinger (1944) was emphatic that the question, What is life? would require essentially *new physics* to deal with it (see chapter 1). On the other hand, Jacques Monod, who was no physicist, vehemently espoused the reductionistic position that any such claim was blatantly vitalistic, as was indeed any suggestion that one could learn anything new about matter (i.e., about physics) from a study of life. In his view, organisms were mere contingencies in a universe governed by "the same laws" whether life was present in it or not. This was his view of the "objectivity" of nature; since organisms and mind are contingencies, their existence could not be predicted from those laws, only discovered. Indeed, in his view, the laws of physics were already completely known and had been known since the nineteenth century.

Let us now turn briefly to what reductionism says about properties such as P_i above, into which we have fractionated our arbitrary property P of a given material system x. The essential assertion here is that each of these is to be associated entirely with, or localized in, some particular subsystem (*fraction*) of x itself. That is, we can partition, or fractionate, any material system x into independent pieces x_j. By independent, we mean (and we understand intuitively) that these pieces can be separated from the larger system x, and from each other, in such a way that their individual properties or predicates $P_k(x_j)$ are (1) entirely independent of any such fractionation process or, indeed, of any larger context whatever, and (2) precisely the individual summands appearing in equation 8.1, the necessary conditions for $P(x)$ to be true, and collectively a sufficient condition.

This presumed *context independence* of the fractionated pieces x_j, into which an arbitrary material system x is to be dissected, is the embodiment of their objectivity, and of the properties or predicates they manifest. Another way of saying this is that the states of x, the totality of true propositions about x, must not contain information coming from any such larger context. Rather, it must pertain entirely to such fractional subsystems. It is precisely this context independence that renders reductionism an entirely syntactic exercise, that gives it its algorithmic character and embodies it in the kinds of lists or programs I described earlier. The underlying idea here is that we must only look

inside, at subsystems, and never outside, at larger systems. To do otherwise is what Monod called vitalism. Parenthetically, reductionism cannot abide the concept of function, for just this reason; it requires this
kind of larger context to give it meaning.

In any case, the essential point here is to require that each $P_i(x)$,
each property or predicate of x, be of the form

$$P_i(x) = P_k(x_j),$$ (8.2)

and, hence, that we can rewrite equation 8.1 as

$$P(x) = \bigvee_{j,k} P_k(x_j).$$ (8.3)

The assertion is that every property or predicate or proposition, of the
form $P(x)$, about a material system x, is expressible in the form of equation 8.3. There is even a canonical fractionation of any such system x
that accomplishes this.

So, in particular, let us suppose that x is a material system we call a
brain, and $P(x)$ is a predicate or property of x we call mind or sentience.
The mind-brain problem, at least in one of its various forms, is to
express $P(x)$ in a form like that of equation 8.3. I repeat, this asserts
that $P(x)$ is true if, and only if, x can be fractionated into context-
independent material subsystems x_i, which in turn fractionate the property P itself according to equation 8.3. Likewise, suppose that x is a
material system we call organism, and that $P(x)$ is the predicate or property of x we call living. Once again, we suppose that we can fractionate x into context-independent subsystems such that equation 8.3
holds.

I emphasize again that equation 8.1 or equation 8.3 constitutes a list
or algorithm expressing the truth of P in terms of a program, something
that can be described to, and executed by, a finite-state "machine." In
turn, such a device is a creature of pure syntax. This provides one way
that *computability* gets inextricably entangled into the foundations of
physics, and into epistemology. Indeed, as I have argued previously
(1988a) and shall do so again here, the presumptions behind equations
8.1 through 8.3 are at root Church's Thesis in another form, now elevated in effect to a basic principle of *physics*. I emphasize also that the

summands in equation 8.1 or 8.3 are determined entirely by cognitive means.

The soul of reductionism is in equation 8.2, which identifies properties or predicates P_i of the original, intact system x with those of certain of its subsystems (fractions) x_i. We thus suppose a set of operators on x, which isolate these subsystems from x, of the following form:

$$F_i(x) = x_i.$$

These fractions, we recall, are required to be context independent, so that their properties are the same in any environment.

It is worth looking a little further at these operators F_i, and at the physical processes they represent. Intuitively, such an operation is presumed to break the constraints, or bonds, that hold the context-independent fractions x_i together within the original, intact, larger system x, but without affecting any of the constraints within the fractions (i.e., that hold the fractions themselves together). In the context of equation 8.3, this means that the bonds in x that the F_i break in liberating these fractions are entirely *irrelevant* to the property $P(x)$. Thus only the constraints *within* the fractions themselves are important. Stated another way, only such interfraction constraints constitute *information* about the property P itself.

The basic feature of these suppositions is that fractionation operations such as F_i are in principle *reversible*. Indeed, the upshot of the preceding discussion is that our original system x is to be regarded as some kind of direct product—e.g.,

$$x = x_1 \otimes x_2 \otimes \ldots \otimes x_M, \tag{8.4}$$

where the symbol $\otimes$ denotes an intrafraction bond or constraint of the kind that is broken by the operators F_i. Equation 8.4 should be compared with equation 8.3, which analogously expresses a *property* $P(x)$ as a corresponding conjunction:

$$P(x) = P_1(x_1) \vee P_2(x_2) \vee \ldots \vee P_M(x_M); \tag{8.5}$$

the two expressions are the same, except that at the level of fractionation of properties P, the conjunction symbol replaces the presumed material bond $\otimes$.

Thus we started with a property P (e.g., sentience) of a material system x (e.g., brain) and fractionated it into a conjunction of cognitively determined subproperties P_i of x. This is the content of equation 8.1. It constitutes a list of conditions, each one necessary for the truth of $P(x)$, and all together sufficient for it—an algorithm. The assertion here is that every property $P(x)$ of x is of this character. The next step is to associate these *subproperties* of $P(x)$ with specific context-independent material subsystems (fractions) x_j of x itself. As we did with equation 8.1, we assume that x itself is a conjunction or product of its fractions, except that material bonds or constraints between these fractions replace the conjunctions that hold the properties P_i together in P. Thus x, as a material system, is expressed as a kind of list or algorithm, for which its fractions are individually necessary, and all together sufficient. And once again, we presume that every material system x is of this form and arises in this way.

These presumptions are entirely syntactical in nature. If they are true, then (1) every property P of any given material system x is associated with an algorithm for determining its truth, and (2) any material system x can be algorithmically generated from a sufficiently extensive population of fractions x_i by purely syntactic means. In other words, analysis of a system x, and synthesis of that system, are inverse operations; likewise, analysis of a system property P, and its synthesis, are likewise inverse operations, falling entirely within a syntactic, algorithmic context, and everything is computable.

On the basis of these ideas, we can imagine studying a property P (e.g., sentience) by synthetic rather than by analytic means. That is, instead of starting with a fixed material system x (e.g., a brain), and trying to express $P(x)$ as a conjunction of cognitively determinable subproperties as in equation 8.1, we might rather imagine assembling systems y from a population of material fractions by establishing constraints between them, as in equation 8.4, until we have enough of them to be sufficient for $P(y)$ to be true. This encapsulates an approach I call *mimetic* (see chapter 6).

To describe this approach, let us begin with the observation that equation 8.4 provides us with an alternate characterization of the *states* of the system x. A state can be expressed as the totality of all properties

$P(x)$ about x, the set of all true propositions $P(x)$. But if x is expressible in the form of equation 8.4, and if every proposition $P(x)$ about x is of the form of equation 8.3 or 8.5, then these same states of x can be built syntactically out of the states of the fractions x_p, without ever looking at x itself, and indeed, without even supposing that x itself exists in any sense in the material world. The identity of these two quite different ways of talking about states of x is a direct consequence of supposing that analysis into fractions, and synthesis from these same fractions, are inverse operations. In *Life Itself*, I pointed out that looking at a state of x as a direct sum of its properties, and looking at it as a direct product of its context-independent fractions, are generically different, and this accordingly constitutes an *equivocation*.

Nevertheless, if all these suppositions are granted, then we can produce a material system x, manifesting any desired predicate or property P, by merely assembling sufficiently many fractions according to equation 8.4. We will thereby obtain systems x that are in some sense *sufficient* for $P(x)$ to be true of that particular system x, but not in general *necessary* for the truth of $P(x)$. All we can say here is that two such systems x, x' built canonically out of fractions via equation 8.4, for which $P(x)$ and $P(x')$ are both true, are *to that extent mimics* of each other, or *simulate* each other. That is, they cannot be distinguished from each other on the basis of the property P alone.

In general, systems x obtained in this synthetic fashion, sufficient for $P(x)$ to be true, but not necessary for that truth, possess in some sense too many properties to constitute a fractionation of P in the reductionist sense of equation 8.1 or equation 8.3. To keep the discussion consistent, we need to add another assumption to all the drastic ones we have already made—namely, that any two systems x, x', which are of the form equation 8.4, and for which $P(x)$, $P(x')$ are both true, possess a common *abstraction*, a "simplest" common subsystem y, such that $P(y)$ is true, and that this subsystem is necessary, not merely sufficient, for that truth. As we shall see, we shall run into disastrous problems with equation 8.4 when we try to express y in this form, and hence in terms of a list or algorithm.

Nevertheless, these mimetic approaches have been much pursued over the years and indeed go back to prehistoric times (where they were expressed in terms of the occult notion of *sympathies* and embodied in "technologies" of "sympathetic magic"). In our own century, the same underlying concepts appear under the rubric "artificial" (as in *artificial intelligence* and *artificial life*). Because they have been so much pursued, and especially since they are regarded as compatible with (and indeed, constituting a synthetic inverse of) reductionistic fractionation, it is worth digressing to discuss them in more detail.

Modern mimetic approaches to mind-brain problems have several roots. The first of them comes from the development of the concept of the neural net, beginning with the two-factor networks of Rashevsky in the early 1930s, and followed by their transcription into Boolean form by McCulloch and Pitts (1943) (see chapter 7). The second comes from foundational questions in mathematics, especially those concerned with consistency, and from notions of formalizability and "effective processes," associated particularly with Russell, Gödel, and Church, and perhaps culminating in these directions with Turing in the later 1930s. The third comes primarily from technology, from switching networks and control theory; the best-known people here are perhaps Shannon (see Shannon and Weaver 1949) and Wiener (see 1961) in the 1940s, and von Neumann into the 1950s (see 1966).

Of particular interest here is von Neumann's argument that, in syntactically assembling systems x of the form of equation 8.4 from fractions x_i in a systematic way, we would come across finite "thresholds" of system size (measured in terms of how many elements appear in a system, and/or how many constraints are necessary to establish between these elements), below which some system property P would be generically absent, and above which the property would be generically present. In a sense, these thresholds (of what von Neumann called complexity, but I think it is more accurate to use the word *complication*) represent phase transitions with respect to the property in question. In particular, von Neumann argued that biological properties (e.g., self-reproduction, the capacity to grow, to evolve, to develop) and properties of mind (e.g., the capacity to learn, to remember, to think, to cognize, to be conscious) were associated with such thresholds of complexity. In other words, he argued that, for any system x satisfying equation 8.4, its complexity constituted at least a generic sufficient condition on x, for $P(x)$ to be true or false.

Such finite thresholds are quite analogous to, say, Reynolds Numbers in fluid dynamics, associated at root with questions of similarity and scale modeling (Rosen 1978b, 1985a). Pushing something across such a threshold, into another similarity class, in this case by adding or subtracting elements and/or intra-element constraints, is itself within the general purview of *bifurcation theory.* Although these ideas start from reductionist-like elements, and share with reductionism a commitment to cognition and to syntax (lists and algorithms), they end up in a way quite incompatible with reductionism as a general scientific strategy. For instance, the intrafraction constraints $\otimes$ are not ignorable any more; they enter into von Neumann's "complexity" in an essential way.

In any case, mimetic approaches are all based on the belief that, given any property P, there is some (syntactic) system x, satisfying equation 8.4, for which $P(x)$ is true, and hence at least sufficient conditions for this truth are embodied somehow in x. A much stronger claim, however, is often made—namely that *every* system satisfies equation 8.4, and thus there exist systems y that are *necessary and sufficient* for the truth of $P(y)$. As we have seen, such a y is an *abstraction* of any other system x satisfying $P(x)$. In a sense, this asserts that every such abstraction y of x is a fraction of x.

Let us now return to the main argument. The common thread that relates reductionistic, analytic approaches to a given material system x, and mimetic approaches, based on synthetically generating other systems y such that $P(y)$ and $P(x)$ are both true, is the presumption of a list or algorithm—that is, a presumption of computability. This presumption is a form of Church's Thesis, expressed in terms of the causal entailments in the material world of physics, rather than the inferential entailments of mathematics. In all other ways, however, reductionism and mimesis are quite different.

However, with regard to properties such as life and mind, there are no such lists that itemize either necessary or sufficient conditions on a system x, or on the property P, for entailing the truth of $P(x)$. In turn, this raises the possibility, at least, that P itself possesses the property of noncomputability, and thus that Church's Thesis is false as a physical assertion about the material, causal world. If it is false, there is no possibility of solving, say, the mind-brain problem, either by reducing mind to brain, or by mimicking mind in "machines," or by any other manifestation of syntax alone.

A purely syntactic world is a very small world, very poor in entailment in general. It has turned out, as shown by generations of attempts to reduce mathematics to syntax via Hilbertian formalizations, that most of mathematics eludes formalization. That is, formalizable mathematical systems are excessively nongeneric in the totality of all mathematical systems. The fact that one can formalize (small) parts of a genuine mathematical system, such as Number Theory, does not in any way constitute evidence that the system itself, as a whole, is formalizable; it is, rather, an artifact. Nevertheless, there is also no way to fractionate Number Theory itself, to partition it into two disjoint parts, one of which is formalizable (objective) and the other of which is not.

Church's Thesis, which originally identified the informal notion of effectiveness with the formal one of computability, accordingly cuts very little ice in mathematics, which generically concerns itself with much larger universes than computability or algorithms or programs allow. Stated another way, abstractions from mathematics that satisfy the procrustean strictures of Church's Thesis are neither necessary nor sufficient for real mathematics; they are, at best, feeble *mimics* of real mathematics.

The failures of formalizability reside, ultimately, in a passion for pure syntax—the complete exclusion of *semantic* aspects. Indeed, these latter are rejected as extraneous subjectivities. The driving idea of formalization is that *semantics can always be replaced by more syntax,* by bigger lists, without changing anything. This idea will be recognized as another form of von Neumann's notion of complexity—that all we need to do is get above some threshold (i.e., add more syntactic elements, and more syntactic rules for manipulating them) and every property will appear, including what seem to be semantic ones.

On the other hand, experience has shown that the resultant lists or algorithms either turn out to be infinitely long, which is unacceptable, or else must turn back on themselves, which is also unacceptable. The latter essentially constitutes an *impredicativity*—Bertrand Russell's vicious circle. Such impredicativities create semantic referents within them, in this case self-referents that depend entirely on the context created by the circle itself. Attempts to eliminate such circles by purely syntactic (context-independent) means, to express them as a finite list, even in a bigger syntactic system, simply destroy these properties.

Since the whole point of formalization is the exclusion of all referents (and especially of self-referents), anything pertaining to an impredicativity has been eliminated from systems governed by syntax alone.

That is, indeed, precisely why they are so small, inferentially weak, and nongeneric among mathematical systems as a whole. Or, to put it another way, real mathematical systems are full of impredicativities— semantic elements that endow these systems with enough inferential structure to close impredicative loops. These, in turn, provide the irreducible semantic aspect inherent in real mathematics itself.

An analogous situation holds in the material world as well, and it is manifested in those material systems we call organisms. Problems such as the mind-body problem appear hard because they involve properties (e.g., life and mind) that depend on impredicativities within the systems within the systems x that manifest them. These are the systems I called complex in *Life Itself* and characterized as possessing nonformalizable (noncomputable) models. This usage, it must be carefully noted, is very different from von Neumann's syntactic employment of the same word.

Just as an attempt to break open an impredicative loop in a mathematical system, and to replace it by a finite syntactic list or algorithm, destroys all properties of the loop itself, so any attempt to fractionate a material system containing closed causal loops destroys all of its properties that depend on such an internal loop. Such fractionations constitute a material version of formalization, artifactual as far as questions like the mind-brain problem are concerned.

All the foundation crises that have plagued mathematics throughout the present century arose from the discovery that set theory was impredicative, that it was itself not formalizable.

Let us conclude with some specific examples that are familiar and that at the same time embody all the difficulties of reductionism and formalization as a general strategy for dealing with things like the mind-brain problem, or the life-organism problem, or the measurement problem in physics.

Suppose we want to determine whether a predicate or proposition $P(x)$ is true for a given x. For instance, suppose we want to determine whether a particular molecule x has a certain active site in it. Or whether a particular pattern x manifests a certain feature. These seem to be nice, objective questions, and we would expect them to have nice, objective, cognitive answers. Basically, these are measurement problems.

The conventional answer is to produce a meter, or feature detector: this is another, different system $y \neq x$, which "recognizes" the predicate P and hence in particular determines the truth or falsity of $P(x)$.

But this recognition constitutes a property or predicate $Q(y)$ of y. We do not know whether $P(x)$ is true until we know that $Q(y)$ is true. Accordingly, we need another system z, different from both x and y, that determines this. That is, z itself must possess a property R, of being a feature-detector detector. So we do not know that $P(x)$ is true until we know whether $Q(y)$ is true, and now we do not know whether $Q(y)$ is true until we know whether $R(z)$ is true. And so on.

We see an unpleasant incipient infinite regress in the process of formation. At each step, we seem to be dealing with objective properties, but each of them pulls us to the next step.

There are several ways out of this situation. The first is somehow to have independent knowledge of what the initial feature is; some kind of extraneous list or template that characterizes what is being recognized at the first stage of the regress; something that makes it unnecessary to cognitively recognize the property $Q(y)$ of y directly, and thus dispenses with all the successive systems z and their predicates R. In other words, we must have some kind of *model* of y, built from this independent knowledge, which will enable us to predict or infer the truth of $Q(y)$, without determining it cognitively. But where does such independent knowledge come from? At least, it comes from outside the system itself, and this would violate the basic tenet of formalization, that there is no such outside.

The only other possibility is to fold this infinite regress back on itself—i.e., to create an impredicativity. That is, to suppose there is some stage N in this infinite regress that allows us to identify the system we require at that stage with one we have already specified at an earlier stage. Suppose, for instance, that $N = 2$, so that, in the preceding notation, we can put $z = x$. Then not only $P(x)$ is true, but also $R(x)$ is true, and not only $R(x)$, but also the infinite number of other predicates arising at all the odd steps in the infinite regress. Likewise, not only $Q(y)$ is true, but also all the predicates arising from all the even steps in that regress.

If, as noted earlier, the states of x and y are defined in terms of the totality of true propositions $P(x)$ about them, then by virtue of what we have said, the impredicativity makes these states infinite objects. In particular, they are objects that cannot satisfy anything like equation 8.1 or 8.3. They also satisfy no condition of fractionability, whereby propositions about x or y can be resolved into independent fractional subsystems by breaking some finite number of intrafractional bonds or constraints. That is, x and y have become *complex* in my sense (1991).

The impredicativity we have described, which is the only alternative to an unacceptable infinite regress, may be described in the following deceptively syntactic-looking form:

$$x \leftrightarrow y. \tag{8.6}$$

This is essentially a *modeling diagram* (1985a) in which the arrows represent what I call *encodings and decodings* of propositions about x into those of y, and which satisfy a property of commutativity (i.e., the path $x \rightarrow y \rightarrow x$ in the diagram is here the identity mapping). But equation 8.6 represents an impredicativity and, hence, from the preceding discussion, it follows that both x and y themselves possess other models that are nonformalizable or noncomputable. Stated otherwise, equation 8.6 actually represents a multitude of inherently semantic *interpretations;* it has meanings that cannot be recaptured from it by purely syntactic means—by lists of other symbols and rules for manipulating them. Moreover, the equation cannot be fractionated into its constituent symbols and arrows by purely syntactic means, without destroying its relation to its own referents. That is, the diagram possesses properties that do not satisfy equation 8.1 or 8.3; it is itself a complex object.

To conclude, let us look at the significance of these remarks, first for the measurement problems of quantum physics, and then for the cognate mind-brain problem itself.

The measurement problem arises because the standard formulations and interpretations of quantum mechanics, which are presumed to hold for a system being measured, do not appear to hold for the new system consisting of the measured system plus an observer. It is a curious problem, which, in the words of d'Espagnat (1976), "is considered as nonexistent or trivial by an impressive body of theoretical physicists and as presenting almost insurmountable difficulties by a somewhat lesser but steadily growing number of their colleagues." The situation is thus somewhat analogous to the reaction of mathematicians and philosophers to the Zeno paradoxes over the years. Nevertheless, an enormous amount of ink has been spilled over it, starting from the early days of the "new" quantum theory, without resolution so far.

In simplest terms, the measurement problem requires a partition to be drawn between an observer and what he observes, so that everything objective (i.e., all the physics) falls to one side of the partition. The trouble is in getting the measurement process itself, as an event in the material world, into that objective side.

Physics, as we have noted, requires as an unwritten law of nature that the presumed objective side with which it claims to exclusively deal be entirely predicative (Church's Thesis), purely syntactic, formalizable. But measurement processes or feature extractions lead either to infinite regresses or to impredicativities somewhere. The presumed partition, which absolutely separates objective physical reality from a subjective observer, has to be drawn so that the requisite impredicativities fall outside the objective part and hence are put entirely into the observer. Indeed, the presumed partition is itself impredicative. Pragmatic physicists claim not to care where the partition is, pointing out that the results of the observations are independent of where they are made.

The parallels between this measurement problem, which tries to objectify the measurement process itself (i.e., reduce it to, and express it entirely within, objective physics), and the mind-brain problem, which likewise tries to reduce the subjectivities of sentience to the objectivities of brain, are clear. They both founder in the face of the impossibility of expressing an impredicativity in purely syntactic, predicative terms. Put another way, there are completely objective properties of material systems that do not satisfy equation 8.1 or 8.3; presuming such stipulations in advance, or restricting ourselves entirely to those properties that satisfy them, and identifying "science" with these alone, is as disastrous a mistake as identifying mathematics with formalizability.

Such questions are already at the forefront in biology. The reductionistic, empirical biologist, who deals entirely with fractionable subsystems, is in the same position as the formalist in mathematics, who deals entirely with formalizable fragments, which are far too small to capture within them the true nature of the larger system from which they came.

The upshot of these considerations is that reduction is not a physical principle, not itself a law of nature. It is violated the instant we venture into the organic realm, where, for example, measurement is. This fact indeed teaches us something new about *physics,* or, more accurately, it forces us to unlearn things we thought we knew about it. If someone wishes to call this vitalism, then so be it. The facts themselves are not changed by the epithets attached to them.

One might justifiably ask, If not reduction, then what? There are several suggestive hints as to the alternatives, already visible in the history of physics and of mathematics. One, to which I have often alluded, is found in the mechano-optical analogy between mechanics and optics, developed by William R. Hamilton in the preceding century. This did not proceed by attempting to reduce optics to mechanics, or vice versa, but established an isomorphism between them based on a homologous action principle. This isomorphism was powerful enough to teach us "new things" about mechanics, to change its very foundations (e.g., via the Schrödinger equation, which was a direct corollary of it) via apparently unrelated optical counterparts. A second consideration arises from what we have already discussed, the relation of a "real" mathematical system to a formalizable fragment of it, and the *absence* of any such isomorphism between them. When judiciously combined, these hints coalesce into an alternate strategy, which has been called *relational.* In a sense, such considerations are an inverse of a process of reduction: the latter seeks to express inherent impredicativities entirely within a purely predicative syntactic framework; the former seek to pull the predicative into a sufficiently rich framework that is impredicative (*complex*) from the outset, and reassemble them therein.

Part III

∾

ON GENERICITY

THE CHAPTERS in this part are loosely organized around the concept of genericity. I devoted chapter 2 of *Life Itself* to this concept,
to what is special and what is general, in connection with the widely
held belief that biology is only a special case of the general laws of contemporary physics (i.e., that biology reduces to physics, as presently understood).

It was a main thrust of *Life Itself* that what is more generic does not
merely reduce to what is less so—it is more the other way around. In
particular, I argued that mechanistic systems are nongeneric in several
basic ways; they are *simple,* whereas most systems are *complex.* I characterized the nongenericity of simple systems in terms of their infinitely
feeble entailment processes, either causal entailment in natural systems,
or inferential entailment in formalisms. From the standpoint of complex systems, simple ones are infinitely degenerate; they are like a high-
order polynomial, most of whose roots coincide.

The most consistent exploitation of genericity, as a primary source
of information, is found in René Thom's work, which began in pure
mathematics in the classification of singularities (for which he received
the Field Medal, the mathematical equivalent of a Nobel Prize). It extended into the world of natural systems, through biology, in his unique
book *Stabilité Structurelle et Morphogénèse,* and, most recently, into the
world of natural language. The essence of his approach is to compare
something, in a systematic way, with neighboring things—those other
things that are, in some sense, nearby. Roughly speaking, a property is
generic if most of these neighboring things possess it. Thus a generic
property is unchanged by an *arbitrary* small perturbation. One of the
empirical challenges to testing genericity in this way is to be sure that

we are applying an *arbitrary* perturbation, one that can lift the degeneracy that is itself typical of a nongeneric, structurally unstable object.

Thus genericity arguments shift attention from a specific thing to what a neighborhood of that thing is like. It is thus a highly context-dependent concept, and one that involves inherent impredicativities. One of my favorite examples arose from the dispute about whether the δ-function, introduced by Heaviside and popularized by Dirac, was a legitimate mathematical object or not (von Neumann, among others, said it was not). It turned out that the δ-function was not only legitimate, it was far more generic than ordinary functions are. Hence, a neighborhood of an ordinary function consists mostly of distributions; ordinary functions are rare and atypical exceptions. On the other hand, if (as was historically the case) we *start* from functions, the only way to get to a distribution is through a transcendental operation, the taking of limits. In other words, what *looks* like an arbitrary perturbation of a function turns out to be very special and only gives us another *function*. We shall see this kind of theme again and again in the chapters of this part; the lesson is that, even though a property is generic, it can be very easy to miss if you start from the wrong presumptions.

Chapter 9 is from an essay that I was invited to contribute to a *Festschrift* for René Thom; it will appear therein, in a French translation. I hark back here to the Schrödinger question, What is life? as an exemplar of an interchange of noun and adjective. Tacit in this question is the idea that this "life" is a thing in itself, and not merely an adjective of other things. Such an idea is commonplace outside biology: people do not think it odd to conceive of, say, turbulence or rigidity as objects of scientific study in their own right, apart from any specific fluid that is turbulent, or any material object that is rigid. I treat complexity in precisely this fashion in *Life Itself*. Nevertheless, in a relentlessly empirical science, which biology has become, one that centers around specimens and deals only with extrapolations and inductions from them, such considerations seem grotesque in the extreme. It immediately puts the discussion outside their reach, and, for perfectly understandable reasons, they do not like that.

Be that as it may, I argue that such an interchange of noun and adjective is tacit in the concept of genericity, with its shift from specific objects to what their neighbors are like.

Chapter 10 continues the theme with an examination of natural languages. This chapter, prepared for a workshop on Language and Reality, concerns the reduction of semantics or meaning to pure syntax, rules of manipulation independent of meaning, and emphasizes the nongenericity of syntax. I introduce a unique instance of the invocation of extra-mathematical referents in purely mathematical contexts—namely, Hadamard's concept of the "well-posed problem" in discussing questions of existence and uniqueness of solutions to differential equations. These external referents, characteristic of natural languages, continually push the boundaries of syntax outward, to larger and larger systems, rather than to the smaller ones that reductionism likes.

Chapter 11 is, on the surface, of a more technical mathematical character than most of the others, and it is one of the few to antedate the publication of *Life Itself*. It is concerned with the concept of genericity, with the exploration of a neighborhood of some object of study by means of a certain family of perturbations of it. The device here, much studied by René Thom, is that of an *unfolding*. The unfolding parameters provide, essentially, a coordinate system that spans a part of the neighborhood in question. The problem is whether this coordinatized part is sufficiently rich to be typical of the whole neighborhood; if so, the unfolding is called *versal* or (if the dimension of this coordinate system is minimal) *universal*.

One prong of this chapter pertains to what I have discussed in the preceding paragraphs—that a neighborhood may be much too big to be properly spanned by unfoldings, even in principle. On my mind here was work I had done earlier, based on patterns of activation and inhibition, of agonism and antagonism, etc., and their relation to ordinary dynamical system theory (as described, e.g., in my *Anticipatory Systems*). This work is couched in the language of differentials and differential forms, not derivatives. And, just as ordinary functions swim in isolation in a sea of distributions, differential equations (exact differentials) swim in a sea of inexact ones. A truly arbitrary perturbation of an exact differential will thus, generically, never preserve exactness. And yet, what looks like an arbitrary perturbation (e.g., an unfolding) of a dynamical system will do just this—it will preserve exactness in the corresponding differential forms. This observation provides the basis for a good deal of still unpublished work, which illuminates notions of complexity as developed from a different viewpoint in *Life Itself*, and also some very practical problems of specificity and recognition that lie at the heart of the concept of control.

In chapter 11, however, I focus on an issue of reductionism. I stress the interpretation of unfoldings in terms of failure modes of structures. This leaves anabolic modes, or creation modes, entirely unaccounted for. Accordingly, a universal unfolding either consigns such creation modes to the nongeneric or else indicates that the neighborhoods we are probing with the unfoldings are too small, and hence that the unfoldings themselves are too small. I also allude to the interpretation of unfoldings in terms of failure modes in terms of the breaking of classical constraints, especially holonomic ones like rigidity. But I do point out explicitly that anabolic (e.g., synthetic) processes are not generically simply the inverse of failure modes, or catabolic (analytic) processes run backward.

Chapter 12 was written at about the same time as its predecessor, and in the same spirit. The target here was the nongenericity of classical thermodynamics, of systems whose equilibria are governed by equations of state such as the Ideal Gas Law, or the van der Waals equation. If there are N state variables involved, then the equation of state uses up $N-1$ dimensions of the state space in unmoving equilibria, leaving only one dimension free for dynamics. I tie this highly degenerate situation to closure and isolation conditions characteristic of classical thermodynamics, the point being the nongenericity of closed, isolated systems in general. This includes, of course, the discrete, context-independent fractions into which reductionistic analysis seeks to partition organisms. I argue from these considerations that closed, isolated systems, as a class, are so degenerate that "opening" them leads to systems that depend much more on how they are opened than what they are like when closed. That is why, I claim, there is today no truly satisfactory physics of open systems, and why they must be studied entirely on dynamical grounds, not thermodynamic ones.

Chapter 13 was written to try to demystify chaos for many of my biological colleagues. I put it in the context of autonomous versus forced dynamics, in the light of profound discoveries about the genericity of "strangeness" in (continuous) dynamical systems generally, and their structural stability. It was not written to attempt to judge, or prejudge, the potential impact of chaos in science—it is far too early for that, in my opinion. The intent was merely to clarify the conceptual issues involved, as I understand them. But the overall theme is genericity, and that is why I believe the chapter belongs here.

༸

CHAPTER 9

Genericity as Information

The appearance of René Thom's *Stabilité Structurelle et Morphogénèse* in 1972 (translated into English in 1976) was a watershed event in mathematics, in theoretical biology, and for the philosophy of science generally. The questions he raised in that book, both directly and by implication, were deeply disquieting to most practicing, empirical scientists, making their dogmatic slumbers untenable. Their predictable responses took several forms: (1) outrage and indignation; (2) violent but irrelevant counterattacks; (3) pretense that Thom did not exist, and hence that his ideas did not need to be addressed at all; and (4) a distortion of his views into a benign soporific that would enable them to sleep again.

Indeed, most of the truly radical aspects of Thom's book have never been discussed at length; the book itself is largely viewed as a wrapper for the Classification Theorem. But that theorem, conspicuous as it is, is only the tip of an enormous iceberg. I will go beneath the water a bit to explore some of the architecture of that iceberg, especially that concerned with the implications of genericity for science in general, for the deeper relations between biology, physics, and mathematics it reveals, and, above all, for the theoretic and philosophical principles buried in it. A prominent part of this discussion will comprise a few of the linguistic aspects to which Thom was increasingly drawn over the past few decades.

I would like to dedicate this chapter to René Thom—a man whose intellect, vision, and, above all, character are all too rare in today's drab world. I hope it will serve to capture some of his spirit.

On Theory in Biology

To someone familiar with mathematics, the word *theory* has no pejorative connotation—quite the contrary. There is no opprobrium attached to those who study group theory, or Number Theory, or the theory of functions of a complex variable. And there is no inherent opposition between those who study such theory and those who study particular instances to which it applies.

In science, however, the situation is quite different. Here, the word *theory* carries an inherent pejorative connotation. Why? Because science lusts for *objectivity,* which associates entirely with (1) the study of *particulars,* and (2) what linguists call *context-independence.* When science is emancipated from "anthropomorphic elements" (see Bergman on Max Planck, at the beginning of chapter 5), only particularities are left. And not even all of them. *Theory* is entirely associated with intrusions of "the structuring mind" (Bergman again); mental constructs superimposed on real, objective, particular facts; contaminants of science, and of the particular, objective data that are alone to be allowed. Hence *theory* connotes the purely speculative, and a false opposition between objective empirics and subjective theory is created, in which the former is good, the latter bad.

Nowhere are such misconstructions more prominent than in biology, especially in the minds of empirical biologists. Modern biologists are simply ashamed of their own intellectual history—almost as an ethnicity they wish desperately to repudiate. They want to be assimilated, as scientists among scientists, but they bear a historical cross in the form of millennia of speculations involving finality, vitalisms, and homunculi of all kinds. They also dread alienation from their own field by the advance of (to them) incomprehensible mathematical developments. This simmers even in physics, usually below the surface but occasionally erupting into open view. A unique, unprecedented, and quite spectacular example was provided earlier in this century, in the form of what was called *Deutsche physik,* or Aryan physics. Although it soon became cloaked in political and racial rhetoric, it was at root a rebellion of empirics against theory, a rebellion of empiricists against the developments in relativity and quantum mechanics, which were incomprehensible to them and hence deeply threatening (see Beyerchen 1977). However, the

only abnormal thing about this particular eruption was its visibility—
no science is immune to it.

Thom's book ran head on into these walls, walls he did not realize
were there, because he came from another tradition. But he quickly discovered that the world of mathematics erects walls of its own. This is,
in large part, based on a fear that concern with worldly phenomena,
with applications, inherently pulls an informal element into the mathematical world—an element that debases concepts of rigor that lie at the
very heart of mathematics. Indeed, just as much as empiricists fear theory as a "vector," pulling alien formal elements into their experimental
paradise, many pure mathematicians fear the same theory as the bearer
of alien informal elements into their mathematical one.

On the other hand, there is at root only one reason to do either
science or mathematics, and that is to understand what is going on. But
understanding is an inherently subjective matter; it is what theory is
about, and it is what theory is for; it is also what makes theory necessary.
The only important thing, as is true for empirics, is that it be done
well. To repudiate it in advance, on the grounds that it is "only theory,"
constitutes a deep, but widespread, intellectual blemish.

On Information

I use the word *information* here in a semantic sense, a sense pertaining to meaning, rather than in a Shannonian, purely syntactic sense
(see Shannon and McCarthy 1956). Roughly, in this usage, *information*
consists of a possible answer to an interrogative, a question. It is interesting to note that interrogatives themselves belong only to natural languages; the symbol ? has no other formal or logical status.

In a natural language, the answer to an interrogative is generally an
adjective of something; it thus comes to modify a noun. This noun is
what the (semantic) information is *about*—what the information *refers
to*. Indeed, the idea of such semantic information without a referent is
vacuous.

I do not specify where such information comes from, but intuitively
we can see that it arises through some process of observation, or through
some process of entailment that involves other information. In what

follows, I shall consider the adjective *generic* both as constituting information about something and as a source of information about other things. That is, I shall be concerned with what kinds of questions, and about what referents, can be answered by the term *generic,* and, moreover, with what other kinds of questions can be answered by means of entailments from this term—that is, with what *generic* can itself imply.

The employment of the term *genericity* as an inferential principle has always played a prominent part in Thom's thinking, first in mathematics and then elsewhere. It also, although in a much different form, underlies the arguments used by Paul Cohen (1966) in his treatment of the independence of the continuum hypothesis and the axiom of choice. In science, things like statistical mechanics could not exist without it. Moreover, the entire concept of *surrogacy,* the replacement of one thing by a different one, without which experimental science would be unthinkable, is closely related to it. Yet as far as I know, the links between genericity and (semantic) information have never been studied systematically.

Genericity in these senses makes the vision of Planck, mandating an inherent, absolute separation of a presumably objective world from the interposition of a "structuring mind," quite meaningless. Stated another way, there is an ineluctable semantic component to objectivity.

On Nouns and Adjectives

Science, especially physics, claims to concern itself exclusively with objective, context-independent things, completely divorced from the subjectivities arising from a "structuring mind." In its simplest form, this asserts that only particular things are real, and that we learn about them through enumeration of their adjectives. Anything else is the province of theory. (Parenthetically, however, we may note that it is adjectives that are directly cognized as phenomena, and the nouns they modify are already products of a "structuring mind.")

According to this view, a *set* of such particulars is not itself a particular, and accordingly is allowed no such objective reality. Therefore, anything based on the formation and properties of such sets is, strictly speaking, outside the province of objective science. In particular, classifications such as the periodic table or Linnaean taxonomy constitute mere subjective intrusions superimposed on the particulars they assort and classify. We learn nothing objective about a chemical element by calling it a halogen, nothing objective about an organism by calling it a vertebrate. Such adjectives, pertaining to sets of things rather than exclusively to the things themselves, convey no *objective* information about the particulars to which they refer.

Indeed, strict empiricists quite rightly feel that admission of such subjective classifications into their science, although of course they do it all the time, is to open a Pandora's box. Once we start forming sets, relations, and classifications on the basis of how particulars look or behave, there is no stopping. We might, for instance, note a basis for relating systems of entirely different character, such as an organism and a dynamical system (i.e., establish a relation between a material and a mathematical system). And even, ultimately, use such a relation to obtain information about the one in terms of the behaviors of the other. From a viewpoint of strict empiricism, such possibilities are simply monstrous.

Let us look at these matters a bit more closely. An empiricist, concerned with particulars such as water, oil, and air, which he will for convenience collect together as "fluids," has no trouble with phrases such as *turbulent water, turbulent oil,* and *turbulent air.* Here, the adjective *turbulent* correctly modifies a particular noun to which our empiricist will grant an objective status, or reality. But suppose we turn these phrases around, to yield *water turbulence, oil turbulence,* and *air turbulence.* We are now treating the *turbulence* as if it were an objective thing, and the water, oil, and air as instantiations or adjectives of it. To treat this *turbulence* as if it were a thing in itself, to be studied apart from any specific instantiation, and indeed, capable of telling us new things about these particulars, is intolerable to the empiricist.

Such an interchange between nouns and adjective provided, for instance, the basis for Schrödinger's question, What is life? (see chapter 1). An orthodox empirical biologist finds such a question meaningless—*life* and *living* are harmless adjectives applied to certain kinds of material systems, conveying information about them. But Schrödinger's question turns everything upside down; it turns life into a thing, and any particular organism into an instantiation or specimen or adjective of life. We could have asked a Schrödinger-like question, What is turbulence? or, What is rigidity? or, more generally, What is X? for an arbitrary adjective X.

Strict empiricists see such considerations as nonsensical at best, hopelessly metaphysical at worst. To grant independent existence to an adjective, to treat it as a noun, involves a mode of analysis they cannot admit. They are willing to partition a noun into other nouns, a process akin to resolving a mixture into pure substances. But the relation between a noun and an adjective is more subtle than that; in particular, it cannot be regarded as a mixture in this sense. There is no *empirical, physical* process capable of making such a separation, and therefore it falls outside experimental analytic modes by its very nature. Nor, for that matter, is there a *synthetic,* inverse procedure for producing a noun that realizes a given adjective (e.g., *living*) by simple juxtapositions, or the creation of mixtures. Indeed, there is something inherently *nongeneric* about what can be done by pursuing such simple-minded, syntax-based strategies, on the grounds that only these must be accorded objective status.

It is the possibility of such interchanges between particular nouns and their adjectives that provides the basis for concepts of *surrogacy.* (See chapter 6, section on Turing Test.)

Genericity

The concept of genericity arises ultimately in attempts to validate *induction* as an inferential procedure. Induction is the procedure of arguing from particularities, or instances, or examples, to the general. In its simplest form, it says that if we know $P(x_1)$, $P(x_2)$, . . ., $P(x_n)$, where P is some property or predicate (i.e., an adjective), and the x_i are samples (nouns) drawn from some larger set X, then we can conclude something like

$$\forall x P(x), \; x \in X. \tag{9.1}$$

If such an inductive inference is valid, then the predicate P is called *generic.* Conversely, if a predicate P is generic, then so are such inductive inferences regarding it.

We see immediately the close relation between the idea of surrogacy (a relation between the x_i) and the genericity of a predicate or adjective that allows surrogates to be interchanged without affecting these relations.

These considerations are generally qualified somewhat to allow (rare) exceptions to expression 9.1. In some sense, the universal quantifier $\forall$ is weakened from "all" to "almost all," say by considering our original set X to be a subset of a larger set Y, such that the difference $Y - X$ is nearly empty according to some measure.

Genericity, in this sense, is a property of properties, an adjective of adjectives. Moreover, it is highly context dependent; it depends very heavily on boundary conditions. Specifically, whether an adjective or predicate P is generic or not depends very sensitively on the set X of underlying nouns that are allowed, and on how its "size" is measured. These will, of course, vary from situation to situation.

Correspondingly, an adjective or predicate P will be called *nongeneric* if its negation is generic. Of course, in most contexts, a predicate P will be neither generic nor nongenetic; that is, genericity is not a generic property of predicates.

Genericity and nongenericity turn out to be closely related, and the following examples demonstrate the implications of this.

1. It is nongeneric for an "arbitrary mathematical system" to be formalizable. This assertion is basically a corollary of the Gödel Incompleteness Theorem, and it expresses the generic presence of semantic aspects in systems of entailment, both inferential and causal—hence, equivalently, the inadequacy of syntax alone as a basis for either mathematics or science (especially biology).

2. Given a suitable family $\{\Phi_a(x)\}$ of parameterized mappings, if a_0 is a bifurcation point, and δ denotes a perturbation of a_0 (necessarily arising *outside* the system), then it is nongeneric for $\delta(a_0)$ and a_0 to parameterize similar mappings. Note that this is a genericity condition on the "perturbation" δ.

We are going to interpret these results in the context of obtaining information, in the sense of answering questions about something, by using information about other, nearby systems, and invoking a concept of surrogacy to justify it. This is a manifestation of what used to be called robustness—what we nowadays call structural stability. When we can do it, it is a manifestation of the genericity of something. When we cannot do it, that very fact reflects the genericity of something else. Both possibilities, as it turns out, are essential sources of information.

A Few Consequences of Genericity

Genericity provides a context that justifies inductive inferences—a way of arguing that, because *this* specimen x has property P, then *all x* will have it. Conversely, if P is generic, and if we are interested in a particular specimen x, we can obtain information about x by interrogating a *different* specimen y—a *surrogate.* In a sense, the more generic a property P in a particular context, the more such surrogates there will be—the more nouns x will satisfy $P(x)$.

Genericity of a property P can be expressed by saying that $P(x)$ implies $P(\delta x)$, where δ is a small but otherwise *arbitrary* operation on, or perturbation of, the particular specimen x in question. That is, δx is always, or at least almost always, a surrogate for x.

On the other hand, we have already noted that genericity of a property P is very sensitive to the context. This shows up in a number of subtle ways, both in mathematics and in science. For instance, it is a tenet of reductionism that there are generic ways of analyzing any material system, and that Newton had already basically discovered them. It is also a tenet of molecular biology that it is nongeneric for a material system to be "alive" (i.e., to be an organism). From these presumptions, it is concluded that (1) biology is only a trivial special case of physics, and (2) we can learn nothing new about physics by studying biology. In fact, all these presumptions are generically mistaken, the conclusions drawn from them spectacularly wrong. These mistakes, woven into the very fabric of contemporary biology, were pointed out by Thom long ago.

Here is one example of how subtle the concept of genericity can be. Consider a system of rate equations, of the form

$$dx_i/dt \;=\; f_i(x_1, \ldots, x_n), \; i \;=\; 1, \ldots, n.$$

The rate functions f_i, or rather their differentials df_i, can be written in an apparently equivalent way, as

$$df_i = \Sigma u_{ij} dx_i,$$

where we have written

$$u_{ij} = \partial f_i / \partial x_j$$

(see my *Anticipatory Systems*). The functions u_{ij} express the activation or inhibition of the growth of x_i by x_j, and hence the differential df_i has a different interpretation from that of f_i, although they seem mathematically equivalent.

To study (structural) stability properties of such systems, we may proceed by "unfolding" the f_i—perturbing them in a presumably arbitrary way by writing

$$\delta f_i = f_i + \epsilon_g,$$

where ϵ is a "small" parameter, and g is arbitrary. This constitutes a generic perturbation of f_i. On the other hand, every such perturbation preserves the exactness of the presumably equivalent differentials. But of course, exactness is a highly nongeneric property of differential forms, certainly not preserved under arbitrary perturbations of the *forms*. Hence, what looks like an unfolding of f does not yield an unfolding of df. Stated another way, there are more ways to perturb df than we can find by perturbing f; df has many more "neighbors" as a form than f has as a function.

This kind of situation is itself generic. There are, typically, many more ways to perturb something than there are things. And then we can think of perturbing the perturbations, and so on. So what is meant by an arbitrary perturbation is a highly nontrivial matter.

These kinds of ideas provide one entrance into the class of systems I call *complex*. See *Life Itself* for another entrance, based on the nongenericity of computability.

The Genericity of Catabolic Modes

Many authors, for example Poston and Stewart (1978), have talked about classical catastrophe theory in terms of failure modes of equilibrium mechanical structures. In these interpretations, a catastrophe is regarded as the breaking of some kind of mechanical *constraint*, such as rigidity. This breaking of constraints is essentially what I mean by a catabolic mode. Correspondingly, an anabolic mode would involve the creation or generation of such constraints.

Both physically (since dissipative or irreversible processes are involved) and mathematically, we cannot in general arrive at anabolic modes simply by running catabolic ones backward. In other words, descriptions of modes of system failure do not, in themselves, yield descriptions of system generation.

Such ideas are basic for biology in a number of ways, as Thom has argued. For one thing, the experimental procedures of biochemistry and molecular biology are almost entirely based on the massive breaking of constraints in a system under study—that is, they begin by making the system fail. We then study the resultant rubble, as if it were a surrogate for the original constrained system. But, of course, pushing a system through a bifurcation in this way means that what we end up with is no longer a surrogate for what we started with.

In another direction, all scenarios for the origin of life presume something equivalent to the genericity of anabolic modes. Thus, in effect, analysis (system failure) and system generation are inverse processes.

Clearly, all these kinds of genericities cannot be invoked at the same time, at least not within the confines of the original system or class of systems we thought we were dealing with. If we want them all, we have only two choices: to try to find a smaller class, a subclass with all these properties, or else to enlarge our class to a bigger one that possesses them.

Speaking generically, only the latter option is viable (no pun intended). It is typically true that the inversion of even syntactic operations such as addition, multiplication, and exponentiation required massive enlargements to permit their inverses to exist generally. And even then, these inversions (e.g., subtraction, division, root extraction) could not be expressed simply by running the direct operations backward. Although it is easy to get the coefficients of a polynomial from its roots, it is not so easy to go the other way.

We have already seen a kind of enlargement that is of the character we need—namely, in passing from a rate function such as f_i, which has only a limited set of neighbors (unfoldings), to a differential df_i, which has a much larger one. Although we cannot pursue the details here, note that it is necessary to gain information about systems by going to larger systems, and not exclusively to smaller ones. Reductionistic modes of system analysis, concerned entirely with objectivity, context-independence, and syntax, cannot abide such information, derived from a larger context instead of a smaller one.

Some Conclusions

I have tried to indicate how the concept of genericity is a powerful source of semantic information, first by discounting the presupposition that a sharp line could be drawn between objective reality and the subjectivities of a "structuring mind." Such a line would leave us only with particularities; it would expunge all classifications and eliminate all notions of surrogacy. I concluded that such a presumed partition has itself no objective existence.

I then characterized *information* in a semantic sense, in terms of interrogatives pertaining to external referents. Thus information becomes associated with propositions $P(x)$ in which the referent x plays a central role.

Next, I argued that the roles of P and x could be interchanged, so that P becomes the referent and x becomes a property or adjective of P. Such an interchange is often called *abstraction* of x. I illustrated such interchanges by means of the familiar adjectives *rigid, turbulent,* and *living.* I indicated how these interchanges provide intrinsic classifications, and surrogacies, into the world of particulars we started with.

I then turned to the concept of genericity, in terms of a context in which it maximizes surrogacy. We saw how it becomes a source of information about a particular y, entailed from a different particular x, using the genericity of P to entail $P(y)$ from $P(x)$.

Most of these considerations are already explicit or tacit in Thom's book of 1972. I hope that my limited remarks will draw attention back to the truly radical facets of that book, and to some of their fundamental implications.

❧

CHAPTER 10

Syntactics and Semantics in Languages

The crucial feature of natural languages is simply that they are *about* something outside the languages themselves. It is their essence to express things about external *referents*. These referents convey meanings on expressions in a language, and, most important, it is entirely by virtue of these meanings that the expressions become *true* or *false*.

The province of meanings is the realm of *semantics*. What is left of a language when the referents are stripped away constitutes *syntax*. Syntactic aspects of a language seem to take the form of a limited number of rules for manipulating symbols and expressions, and they look deceptively easy to formalize. Moreover, syntax appears, partly for that very reason, refreshingly objective and hence amenable to *scientific* scrutiny, constituting a closed and tidy universe we need never leave. Semantic aspects, by contrast, possess ineluctable subjective features, which often appear entirely arbitrary, idiosyncratic, and conventional—stigmatized with the epithet *informal*, poor material for scientific investigation.

Accordingly, a centuries-old trend has developed, based on the idea of reducing semantics to syntax. The idea is, roughly, that the more syntax there is in a given language, and the less semantics, the better (i.e., the more objective) that language will be. Thus the best language will have no overt semantic aspects at all. The strategy is that specific meanings, coming from external referents, can be *equivalently* replaced by suitably enlarging the set of syntactic rules for manipulating expressions within the language; the germ of such a strategy goes back to the *Characteristica Universalis* of Gottfried Wilhelm von Leibniz, among others.

One of the main sources of referents, meanings, and truths expressible in language is what I have called the *external world*—the material world of cognizable events, which we observe and measure, and with which science deals. One of the profound consequences of a *Characteristica Universalis* is its claim that, in principle, we could dispense entirely with this world and subsume its laws into syntax alone. The most modern expression of this belief is Church's Thesis (see Rosen 1962a), which projects back from language into an exceptionally strong restriction on the nature of that external world. The question is whether or not the thesis itself is true or not—that is, whether semantics can always be equivalently reexpressed in terms of more syntax. This chapter is mainly devoted to the consequences, both for language and for science, of its falsity.

One area in which this strategy has been most relentlessly pursued is in a branch of language called mathematics. The ongoing quest for axiomatization is a form of this trend, going back to Euclid and before. In our century, in large part as a response to the foundation crises arising in Set Theory, this quest took a variety of forms, one of which was the formalism of David Hilbert. Hilbert was led to deny that (pure) mathematics was *about* anything at all—even about itself. Indeed, he implicitly blamed the crises themselves entirely on unexpunged "informal" semantic residues in mathematics, and he proposed replacing the entire enterprise with an inherently meaningless game of pattern generation, played with symbols on paper. Within a short time, however, Gödel proved his celebrated Incompleteness Theorem, which in effect showed that syntactic rules captured only an infinitesimal part of "real" mathematics—in effect, that Church's Thesis was false, even in this realm. Or, stated another way, that mathematical systems that are formalizable, in the Hilbertian sense, are infinitely rare (nongeneric) among mathematical systems in general. So, in this realm, there is no way to reduce semantic aspects to syntactic ones in general; the result is only a mutilation, with the creation of artifacts.

Of course, it may be argued that the mathematical experience is itself somehow atypical, and that the external world of material phenomena will behave differently. I propose to investigate a few ramifications of this possibility, especially in the light of the organic realm.

Models and Modeling

An essential part of any language resides in its implication structure. This comprises a system of *entailments,* which generally are expressed in the terse form $P \rightarrow Q$. The main property of such implications, or entailments, is that they propagate "truth" hereditarily across them; thus if P is assumed to be "true" (whatever truth may be, here), then Q must be true also.

It is often supposed that such inferential entailments are part of the syntax of the language and do not themselves depend on any external referents or meanings; only the truth values arise semantically. Of course, such truth values may simply be *posited,* but in either case, they come from outside the syntax itself. We shall not make this supposition in advance; we leave open the possibility that the entailment structure of the language itself can change by virtue of what the language is about—that is, that it can depend on semantic as well as syntactic features. This kind of possibility is denied by the claim that semantics can always be reduced to syntax.

In any event, entailment structures (e.g., as they arise in different languages) can be *compared,* in much the same way as, for example, geometric figures can. That is, there is a notion of *congruence* that can be established between different entailment patterns. (Such notions are examples of *functors* in the Theory of Categories [see *Life Itself,* section 5L].) The bringing of entailment patterns into congruence is the essence of what I have called *modeling relations* (Rosen 1985a, 1991).

As far as science is concerned, it is inseparable as a human activity from the belief that the external world, of material events and phenomena, itself comprises a system of entailments. That is, the events and phenomena that we observe and cognize are not entirely arbitrary or whimsical but must satisfy definite relations. Moreover, these relations may be discerned and articulated, and they collectively constitute the province of *causality* and manifest the system of *causal entailments* described originally by Aristotle two millennia ago. In particular, we can think of comparing a system of causal entailments (roughly, a family of events in a material system) with a corresponding system of inferential entailments, between propositions describing these events.

A congruence between a system of causal entailments (pertaining to what is happening in the external world) and a system of inferential ones (pertaining to a language in which these events are expressed) can be represented as a diagram or pictograph of the form

DECODING

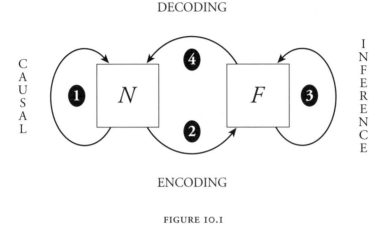

FIGURE 10.1

The left-hand side of this diagram comprises the system being modeled (often called the *object system*). The arrow 1 represents the internal entailments that characterize this object system. The right-hand side of the diagram is a linguistic system, generally a mathematical one, with its own arrow 3 of internal, inferential entailments. The crucial ingredients are the arrows 2 and 4, which I call *encoding* and *decoding*, respectively. (I have discussed the anomalous features of these arrows in more detail in *Life Itself,* section 3H.) They do not fit entirely inside either the object system or the model; they do not represent entailments, nor are they themselves entailed. They manifest what Einstein (with Infeld; 1938:33) once called "free creations of the human mind," on which he believed science depends. They introduce an obvious further semantic element into the model, over and above what semantic (e.g., nonformalizable) features may already be present in the model. The only condition on them is that they bring the two entailment structures into congruence—that is, that they satisfy the commutativity condition, which I have written as

$$1 = 2 + 3 + 4.$$

In the discussions in *Life Itself,* my purpose was to use modeling relations to find out about the object system (the totality of models that can be brought into congruence with an object system comprise its *epistemology*). However, in figure 10.1, the modeling relations are completely symmetrical and therefore can probe the *language* just as effectively. Indeed, by writing the commutativity of the diagram as

$$3 = 4 + 1 + 2,$$

we effectively interchange the roles of the object system and the model.

Accordingly, just as we view the totality of models we may make of an object system as the fullest expression of its nature, we may equally view the totality of object systems that may be encoded into, or decoded from, a given language so that commutativity holds, as a probe of that language. This is, by its very nature, an inherently semantic probe, and it measures what we might call the language's *fahigkeit*—its capability to express meanings.

Indeed, we shall take the view that material object-systems, as things in the external world, and whose behaviors are governed by webs of *causal* entailments, constitute the true province of *effective* processes. That is, the notion of effectiveness has to get imported into language via modeling relations arising in material nature, through encodings/decodings from these. Accordingly, any attempt to characterize effectiveness independent of these, from within language (e.g., mathematics) alone, and, most particularly, in terms of its syntax alone, is perilous. But that is exactly the substance of propositions such as Church's Thesis—namely, that causal entailments in the external world must conform to some notion of effectiveness drawn from inside the language itself. However, it is the other way around.

On Entailment Structures in General

Any system of entailments can be probed exactly as Aristotle proposed probing causality in the material world so long ago—namely, by asking the (informal) question "Why?" Basically, the question "Why x?" is the question "What entails x?"

A system that is weak in entailment is one in which very few such *why* questions about members of the system *possess answers in the system*—that is, most of the things in such a system are *unentailed* within the system. Accordingly, they must be separately posited, or hypothesized, from outside. Thus a good measure of the strength of an entailment system is precisely in how many *why* questions about its members can be answered within the system, and how many are dependent on positing or hypothesizing.

In this sense, syntactic or algorithmic systems, or formalizable systems, are extremely weak in entailment: almost everything about them must be posited, and very little is actually entailed within such a system. That is, almost every *why* question about what is in the system cannot be answered within the system, and thus their answers take us outside the system. In this sense, the objective appearance of formalizable systems, their apparent rigor, is an illusion.

In the fourteenth century, William of Occam proposed as a principle what has become known as Occam's razor: Thou shalt not multiply hypotheses. It is thus a characteristic of inferentially weak systems that thou *must* multiply hypotheses, precisely to compensate for the inferential weakness, the inability to answer *why* questions about the system within the system.

It is a characteristic of strong systems of entailment that they possess enough entailments to allow closed loops. In inferential systems, such loops are often called *impredicativities* (Bertrand Russell's vicious circles). From a purely syntactic point of view, these are bad, because they create an inherently semantic element within the system; indeed, it was the whole point of formalization, or the replacement of semantics by syntax, to do away with them. The result was a far weaker system inferentially, with much less entailment and much more that had to be posited.

In a certain sense, causal entailment structures in the material world are regarded as requiring no positing at all; every *why* question about a material system possesses a material answer. Indeed, the very concept of positing makes little sense in the context of causal entailment. Of course, the answer to a *why* question about a given material system may not be found within that system itself (unless the system has been forever closed and isolated). However, in *Life Itself* I argue in that the systems we call *organisms* are, in a sense, maximal in their entailment structures; asking *why* about them can be answered in terms of something else about them. They thus are inherently semantic and contain causal counterparts of impredicative loops.

Attaching semantic meanings to linguistic objects, as in comparing entailment structures via modeling relations, allows us to import, in a certain sense, some of the presumed richness of causal entailments into language itself. We shall now turn to a specific situation that vividly illustrates some of these issues.

"Well-Posed" Problems

Pure mathematics is notorious for its disdain of external, or at least of extramathematical, referents, although it does not go as far as Hilbert in its rejection of all semantic aspects within mathematics. Any attachment of such external, material referents, as in figure 10.1, is relegated to the domain of applied mathematics, and the aim in such applications is to enrich our understanding of causal entailments by supplementing them with the inferential procedures of mathematics itself.

Actual examples going the other way, in which a diagram such as figure 10.1 is used to enrich our understanding of entailment procedures within mathematics by attaching semantic external referents to purely mathematical entities (i.e., in which applications flow from the external world into mathematics, rather than the reverse), are relatively rare. This is not surprising, given the attitude of most pure mathematicians about "rigor." It is not unusual for pure mathematicians to informally utilize ideas pertaining to causal entailments in their private thoughts (e.g., it is well known that Bernhard Riemann [see Maurin 1997] arrived at many of his analytical insights by thinking of electricity flowing in thin sheets), but these external semantic referents are invariably regarded as extraneous and irrelevant, and expunged before publication.

One familiar instance in which this is not possible, where the external referents are essential, is in the concept of the "well-posed problem" as initially described by Hadamard. In his own words,

> A sure guide is found in physical interpretation: an analytic problem *always* being correctly set when it is the translation of some mechanical or physical question.   (1923)

This idea of the well-posed problem, which tacitly uses a modeling diagram like figure 10.1 to learn something about inferential entailment (arrow 3 in the diagram) from causal entailments (arrow 1) between external referents, has played a fundamental role in dealing with classic mathematical problems, like the Cauchy and Dirichlet problems, more tacit in things like the Plateau Problem (minimal surfaces), or, generally, where questions of existence and uniqueness are involved.

In a sense, Hadamard's concept of the well-posed problem asserts the *mathematical* validity, even the necessity, in analysis, of a form of "analog computation" (as embodied in figure 10.1) to supplement the entailment processes available from within mathematics itself. In turn, these ideas depend in an essential way on the semantic features arising from the involvement of external referents via encodings and decodings. That is, questions of *mathematical* truth or falsity, like the existence and uniqueness of solutions of analytic problems, devolve ultimately on truth-values arising outside those problems.

What are we to make of things like the property of an analytic situation being well posed? This seems to me a perfectly objective property of a purely mathematical entity, like a partial differential equation. But it is a semantic property, one that depends on external referents. It is actually a property of an entire modeling relation, like figure 10.1, of which the mathematical entity is only a part.

The question is, are there means, entirely within mathematics, for deciding whether a given problem is well posed (i.e., whether the given situation possesses this property of well-posedness) or not? And further, even more stringently, whether such decision procedures are, or can be, of a syntactic (formalizable) nature or not? In a certain sense, we are asking whether the property of being well posed is itself well posed.

We can see that questions of this kind raise some fundamental issues about both language (especially about the nature of inferential entailment within it) and material nature (the character of causal entailment between events), depending on which way we chase through modeling diagrams like figure 10.1. Especially, they probe how far a language may be fractionated from external referents at all and still remain a language. This last question is closely related to the balance between semantics and syntax within the language.

Church's Thesis

Church's Thesis was originally offered as a rigorous formalization of an informal semantic concept (that of effectiveness; for reviews, see Rosen 1962a, 1985a). It specifically identified what was "effective" with what was mechanical—more specifically, with software or programs to finite-state machines (Turing machines), themselves considered initially as mathematical objects. But it is very easy and tempting to equivocate on the term *machine,* as for instance in the following citation from an early textbook on the subject:

> For how can we ever exclude the possibility of our being presented, some day (perhaps by some extraterrestrial visitors), with a (perhaps extremely complex) device or "oracle" that "computes" a noncomputable function? (Davis 1958)

It is plain that the thesis, which originally proposed a purely mathematical identification of effective calculability with computability, is here unconsciously elevated to a limitation on material nature itself—namely, that every material system (e.g., the "oracle") can only have models that, considered as purely mathematical objects, must satisfy the stipulation of computability (formalizability). That exceedingly strong stipulation on *causal* entailment, which amounts to mandating its conformity with syntax, locks us into a category of systems I call *simple systems.*

If this is accepted, then computability becomes a sufficient condition that a mathematical problem be well posed, in the sense of the preceding section. Indeed, it can be argued that computability becomes both necessary *and* sufficient—a purely mathematical (indeed, a purely syntactic) criterion for what Hadamard called the correct setting of any analytic problem.

The thesis, via its reliance on machines, also elevates a distinction between hardware (the machine's "reading head," and the locus of all its entailment) and software (its programs and inputs, inscribed on its tape) into a corresponding absolute distinction inherent in the natural world. The stipulation that the machine outputs (software) must not alter the hardware is, on the one hand, a guarantee against impredicativity (i.e., a guarantee of inferential weakness), and, in the material world, the basis of such things as the Central Dogma (Judson 1979) of molecular biology.

The mathematical content of Church's Thesis has to do with the evaluation of mappings or operations—that is, with expressions of the form $Av = w$, where A is an operator of some kind, v is a particular argument, and w is the corresponding value. Church himself was concerned entirely with this evaluation process, and in particular with those he deemed "effective." In very broad terms, he tried to identify this informal notion of effectiveness with the existence of an *algorithm,* a mechanical process, a computation, for the production of a value Av from an argument v. This is how machines (the Turing machines) became involved in the issue in the first place—via the identification of a *program* (software) for A with this evaluation procedure—and, in this process, via the equivocation we have mentioned on the word *machine,* how the thesis came to be imported into science and epistemology, through the idea that every model of every physical process had to be formalizable. That is, such models are purely syntactic objects; nothing semantic is required in them or of them.

That systems (mathematical or material) satisfying the strictures of Church's Thesis are preternaturally weak in entailment can be seen in two distinct (but closely related) ways. The first way is to note that, in an expression of the form $Av = w$, the only thing one can ask *why* about is the value w. The answers are "Because A" (formal plus efficient cause; hardware plus program in a machine) and "Because v" (material cause; input to a machine). Everything else is *posited* from outside—unentailed. One cannot ask "Why A?" or "Why v?" and expect answers within the system.

The second, related way is to note the difficulty of solving *inverse* problems—that is, solving the very same relation $Av = w$ for v, given A and w. This problem essentially requires the production of a new operator, which we might call A^{-1}, and then evaluating (computing) that. But how is this inverse operation itself to be entailed, let alone evaluated? It is well known that these inverse problems cannot be solved from within a fixed syntactic universe. They require an ineluctible *semantic* aspect; indeed, it was precisely to guarantee enough entailment to solve such inverse problems that the concept of well-posedness was developed in the first place, through the invocation of causal entailment via modeling relations.

It may also be noted that a description of something through what it entails, rather than exclusively through what entails it, is the hallmark of Aristotelian *final causation.* Thus, in the above equation, the solvability of its inverse problem lets us answer the question "Why *v*?" by saying "Because *v* entails *w*." This is one of the symptoms of *impredicativity* in the system itself, and it is precisely what syntax alone cannot accommodate, what in fact it was invoked to avoid.

We thus find the following peculiar situation in purely syntactic systems: The more forward operations *A* in the system there are (i.e., the more evaluations or computations $Av = w$ one can perform), the fewer inverse problems one can solve within the system. Indeed, solving the latter requires entailing inverse operators (e.g., A^{-1}) within the system, making them *values* of other arguments and operations already in the system. It is of great interest to note that precisely this capacity, which is necessarily absent in all syntactic systems, is also the hallmark of the class of systems that I have termed (M, R)-systems, the simplest *relational* models of organisms (see *Life Itself*).

Thus, in a way, Church's Thesis plays a kind of antithetical role to the notion of the well-posed problem. They both rely on invocation of modeling relations, but the former does so to limit the content of the mathematical side of these relations, whereas the latter does so to enlarge that content. And modeling relations are inherently semantic, in that they associate external referents to mathematical objects, subject to the commutativity of the modeling diagrams (see figure 10.1). This discussion will conclude, in the next chapter, with some of the implications of these considerations for languages—in particular, with some of the consequences of trying to fractionate languages into syntax plus semantics, and, even more, of trying to reduce the latter to the former.

Some General Conclusions

In this section, I look more closely at the claims that (1) syntax can be fractionated from semantics without changing either, and (2) that semantic aspects, involving external referents, can be reduced to syntax just by adding more syntactic rules. I invoke both the idea of the well-posed problem, which extracts mathematical information from just such external referents, and Church's Thesis, which says that everything is formalizable and hence can be completely separated from all external referents.

The first of these claims, that a line can be drawn between objective syntax and subjective semantics, is quite analogous to many other such claims. The same kind of claim is at the root of the measurement problem, pertaining to objective measured systems and their observers, with the former entirely to one side of a line and the latter entirely on the other (Rosen 1993). Such a partition is involved in a fashionable characterization of physics as attributed to Max Planck by Bergman (1973; see chapter 5). In mathematics, this kind of line is contemplated in Church's Thesis and its homologues, between what is effective (i.e., objective) and what is not. Such partitions form the substrate for the mind-brain problem and the life-organism problem, among many others. It is presumed that the contemplated partition is itself objective.

The second claim has to do with getting across this line. To go from the objective side to the subjective one is the substance of science and epistemology; it involves comprehension, understanding, explanation. Crossing the line in the opposite direction is nowadays essentially reductionism; in our present situation, it is the belief that semantic aspects of a language, on the subjective side of the line, can be equivalently replaced by more syntax on the objective side.

The matter can be viewed in a more restricted context by drawing this line through a modeling relation (see figure 10.1). The main question is what the line would look like, and how it would be expressed, from *inside* the diagram. In particular, what does the line do to properties of the entire diagram, such as its commutativity?

Let us begin by noting that a partition of a universe of discourse into two parts (e.g., a world of propositions into true ones and false ones) creates the basis for an impredicativity. For any proposition pertaining to that partition itself, for which that *partition* is a referent, belongs to that world and, accordingly, must fall into one class or the other. Here is the "vicious circle" at the root of all the classic paradoxes, while at the same time, it is an essential part of classical real and complex analysis, much of topology (especially metric topology), and of measure theory and hence stochastics. That is, at root, the main reason these parts of mathematics are simply *unformalizable;* they can be neither described to a machine as software nor generated by such a machine from software. Impredicativities are simply part of the semantic legacy of mathematics as a language; they express a heritage of *transcendental* operations, like the taking of limits. These, in turn, can only be expressed syntactically in terms of infinitely long propositions, and the assignment of truth-values to these. But restrictions such as Church's Thesis do not like infinitely long programs, or infinitely large state sets in the reading heads. So, as far as syntax is concerned, these infinite objects cannot come from "inside." Rather, they have to come from the *referents,* for which the syntax has become not a language but rather only a *shorthand.*

Indeed, it is quite easy to see that Church's Thesis lends itself to a version of Zeno's old paradoxes. As we have seen, the identification of an "effective" process with algorithms and computability relies on the creation of an external referent—a machine that executes the process. In turn, this creates a modeling relation; a diagram such as figure 10.1, between that machine and the purely syntactic language in which the process is expressed. Between the two, there are definite encoding and decoding arrows, which make what happens in the machine commute with the effective inferential processes in the language.

Now the very commutativity of the diagram itself creates a new kind of inferential process, which we can see as follows. The process of encoding a referent into a language creates a pair $[r, E(r)]$, where r is the original referent, outside the language, and $E(r)$ is its encoding within the language. Commutativity refers to these pairs: roughly, if r entails r', say causally, then $E(r)$ must entail $E(r')$ inferentially. To express this commutativity within the language, we must encode these pairs separately; they are not encoded by the original E. Let us denote the encoding process, which transduces these pairs, by E'. Then the commutativity of the diagram can itself be expressed by an (inferential) entailment, within the language, of the form

$$E'[r, \; E(r)] \text{ entails } E'[r', \; E(r')].$$

By Church's Thesis, this new inferential entailment must itself be "effective," hence computable. It refers to a model of the original modeling process; commutativity of a diagram such as figure 10.1 is a property of the whole diagram, and not of any individual fragment of it.

But this procedure can be iterated. We can do the same thing with triples of the form $\{r, E(r), E'[r, E(r)]\}$, which are again related by a commutative diagram. To express that commutativity in linguistic terms requires a new encoding E'' of these triples, and a new inferential entailment between these encoded triples. And so on. Thus we generate an infinite regress in which, at each stage, we require a new metalanguage to express the commutativity of the procedure mandated by invoking Church's Thesis at the preceding stage.

None of the ways of breaking off this infinite regress satisfies Church's Thesis. It will not do, as is customarily believed, to simply suppose $E = E' = E'' \ldots$, because the referents at each stage are all different, and *referent* is an inherently semantic concept. More generally, to suppose that $E = E^{(n)}$ at some stage n simply creates an impredicativity, which again is a semantic concept. Thus we must either live with an infinite syntactic regress or admit semantic aspects *into the language.* Of course, natural languages always do the latter, because, as noted at the outset, it is their function to be about referents outside the language itself.

A number of conclusions can be drawn from these very simple considerations. One of them is that a language cannot be fractionated into a purely syntactic part and a complementary semantic part. That is, there is no line that can be drawn with all the syntactics falling to one side of it and all the semantics on the other. There is, accordingly, no line that can be drawn between a part of the language that is independent of external referents, and a complementary part that can, or does, pertain to such referents. Stated another way, we cannot identify pure syntax with *objective,* and dismiss all the rest as *subjective.* A language fractionated from all its referents is perhaps *something,* but whatever it is, it is neither a language nor a model of a language.

In particular, if we try to fractionate a modeling diagram in such a way, with all the referents apparently left to one side and all the encodings/decodings put on the other, we lose the commutativity that is the essential feature of the diagram as a whole. Instead, I argue that, if one wants to fractionate, one must do so at another level, at the level of such diagrams or modeling relations. If one tries to fractionate the diagrams themselves, their basic properties are irretrievably lost.

The same kinds of considerations on the failure of reducing semantics to syntax also bear heavily on the validity of reductionism as a strategy for probing natural systems in general. This is particularly true of the endeavor to reduce biology to contemporary physics (see *Life Itself*). The main issue, however, is that, just as syntax fails to be the general, with semantics as a special case, so too do the purely predicative presuppositions underlying contemporary physics fail to be general enough to subsume biology as just a special case. This kind of attitude is widely regarded as vitalistic in this context. But it is no more so than the nonformalizability of Number Theory. Stated another way, contemporary physics does not yet provide enough to allow the problems of life to be well posed.

༄

CHAPTER II

How Universal Is a Universal Unfolding?*

The primary distinction between "pure" and "applied" mathematics is that the latter is overtly directed toward extramathematical referents. That is, in employing mathematics as a tool to study other things, it is not enough to prove theorems; it is at least equally important that the conditions imposed by these extramathematical referents be respected. Such conditions are, by their very nature, informal; that is why mathematical modeling is an art and is in many ways harder than mathematics itself.

For example, ecological population dynamics becomes mathematical ecology when it is expressed as some kind of mass-action law imposed on a state space coordinatized by population sizes or population densities. In this case, the image of a biological population is a dynamical system acting on a manifold of states.

One of the natural things to study in this context is the stability of a biological population, expressed in terms of the stability of its image as a dynamical system. One obvious facet of population stability involves the persistence or extinction of species in the population. To study extinction in this context means to investigate the properties of the imaging dynamics near the coordinate hyperplanes. The literature is full of such studies, involving exemplary mathematical investigations, but without regard for the fact that, because biological populations are quantized, there is no meaningful dynamical referent to behavior near these hyperplanes. Such studies, then, show only that very good mathematics can be combined with very poor modeling.

---

*This chapter previously appeared in the journal *Applied Mathematical Letters* 1, no. 2 (1988): 105–107. Reprinted by permission.

As another example, we may consider the Keller-Segel models for slime mold aggregation. This is a simple system of coupled partial differential equations which determine the density of a population of free slime-mold cells on a bounded substrate. It was shown that, under certain plausible conditions, uniform or constant-density distributions could become unstable. Departure from homogeneity means aggregation, and it was here that the authors stopped.

Subsequently, other authors reconsidered the Keller-Segel model, concentrating on the asymptotic behavior of its solutions. It took a considerable amount of time and effort to show that these solutions made the cells pile up with infinite density at discrete points of the substrate. It took further effort to revise the original equations to avoid this unacceptable behavior. All of this was done despite the fact that the Keller-Segel system stops being a model of aggregation long before we need concern ourselves with asymptotics.

The number of such examples exhibiting a mismatch between mathematical structures and their intended external referents can unfortunately be shown to be very large.

One area of pure mathematics for which external referents, especially biological referents, are being avidly sought is the area of structural stability (particularly bifurcation theory and catastrophe theory [Rosen 1981]). It is our aim to show that also in this area, failure to respect the requirements of such external referents creates the same kinds of problems described above. In such cases, however, it turns out that these very failures can themselves be illuminating if properly interpreted.

As we pointed out long ago (Rosen 1978b), structural stability or bifurcation theory is essentially the study of an equivalence relation (similarity) on a metric space. The similarity creates a taxonomy or classification of the elements of the space, while the metric allows a notion of approximation, or alternatively of perturbation, of the elements of the space. One basic question in this context is to determine conditions under which a whole neighborhood of an element of the space lies within an equivalence class. In such a case, any sufficiently good approximation to, or any sufficiently small perturbation of, the given element is also similar to it. Such a situation is called stable. Otherwise, there must be several (perhaps infinitely many) equivalence classes of the similarity relation which intersect any neighborhood of the element; this is bifurcation. In this situation, the problem becomes to enumerate all these similarity classes (or at least, those which can be reached generically).

The most interesting habitat for such studies are spaces S of mappings $f : X \to X$, where X is a manifold. There is no preferred way to specify such a space S; they mostly consist of differential or analytic functions, whose Taylor series possess particular properties (chosen mainly from considerations of mathematical tractability). If $f \in S$, then a small neighborhood of f consists of all functions of the form $f + \epsilon g$, where $g \in S$ and ϵ is a "small" parameter. Alternatively, we can take the Cartesian product

$$\bar{S} = s \times U_r,$$

where U_r is a subset of r-dimensional Euclidean space containing the origin ω. U_r is variously called a parameter space or control space. Let us consider a mapping $F : \bar{S} \to S$, with the property that $F(s, \omega) = f(s)$. In fact, for every $u \in U_r$, we can define a mapping $f_u : S \to S$, and under suitable restrictions, these mappings f_u populate small neighborhoods of f. Under these circumstances, we can call F an *unfolding* of f. If f is not stable, the problem is whether there are "enough" functions f_u to stably put a representative into every similarity class which intersects arbitrary small neighborhoods of f. This turns out to depend on the dimension r of U_r. If there are enough such functions (i.e., if r is big enough), the unfolding F is called *versal;* if r is minimal, the unfolding is called *universal.*

There are many different interpretations (realizations) of these mathematical objects. One of them involves the "failure modes" of beams, struts, and other kinds of equilibrium mechanical structures. In this case, the mathematical functions f can specify either some property of the equilibrium structure itself (e.g., its shape) or else the dynamics which generates the equilibrium. The parameter space U_r in this case is generally interpreted in terms of the stresses or forcings impressed on the structure.

If such a structure has dissimilar ones near it, then an appropriate small perturbation can push it into a different similarity class, and we may say that the structure has failed (e.g., buckled, broken, collapsed, etc.). Thus every such nearby similarity class gives rise to a different failure mode. Since by definition the universal unfolding touches almost all nearby similarity classes, it is supposed to represent a complete enumeration of all possible failure modes.

The difficulty with this interpretation is that, in general, mechanical failure modes are irreversible. Otherwise, we could rebuild or repair a failed structure by simply reversing a path through U_r which takes us from the original structure to the failed one. On the face of it, however, bifurcation theory should be as much a representation of system generation as of system failure. Indeed, in generating or manufacturing a mechanical structure, such as a bridge, we need the bridge itself to be a bifurcation point, under the very same similarity relation pertaining to failure modes; otherwise we simply could not even build it.

The fact that we cannot in general interpret universal unfoldings in terms of generation modes as we can in terms of failure modes raises a number of deep and interesting questions. Obviously, it implies that the universal unfoldings are indeed far from universal. It might happen, for instance, that construction modes are more special, or less generic, than failure modes, and thus fall outside the scope of the standard types of bifurcation analysis (Elsasser 1958). Or it might happen that we must extend f to a larger manifold, in which construction modes become generic.

At a deeper level, the brief discussion we have given above throws new doubt on the entire reductionist program in biology—a program which is doubtful on many other grounds as well (Rosen 1985a). Empirically, the reductionist program proceeds from the assumption that properties of failure modes completely characterize the function and generation of organisms. Indeed, the empirical approach to biological organization is precisely to make them fail in a variety of drastic ways, and then employ the properties of the resultant rubble to generate properties of the intact organism. The general irreversibility of failure modes, together with the fundamental and inherent dissimilarity between failed structures and those from which they arise, indicates that such a simpleminded strategy is most unlikely to work. Indeed, what we study in such a strategy is more a function of how the failure mode was induced than of the system to which it is applied.

CHAPTER 12

∾

System Closure and Dynamical Degeneracy*

A property of a mathematical object is often called *generic* if a sufficiently small but otherwise arbitrary perturbation of the object produces another object with the same property. In other words, if something is generic with respect to a given property, it is to that extent indistinguishable from any of its immediate neighbors.

For instance, let us consider the set of $n \times n$ real matrices. The property of invertibility (i.e., nonvanishing of the determinant, and hence of any of the eigenvalues) is generic; singularity is not. Possession of real or complex eigenvalues is generic; possession of pure imaginary eigenvalues is not. Distinctness of eigenvalues is a generic property; that of coincident eigenvalues is not. More generally, a matrix whose eigenvalues satisfy any kind of identical relation (so that one or more of them are functions of the others) is to that extent nongeneric. From these examples, we can see that nongenericity is associated with degeneracy, an absence of independence and a lowering of dimensionality.

For many purposes, one wants to study generic properties and the objects which manifest them. However, a great deal of mathematics is concerned rather with objects manifesting special features, satisfying special conditions, or obeying particular relations and identities. Such objects are endowed thereby with the rich possibilities which make for interesting and deep mathematics. But ironically, these very properties are also associated with nongenericity and degeneracy.

When we use mathematical language to image the material world, we tend to deal with the nongeneric. Conservation laws, symmetry

*This chapter previously appeared in the journal *Mathematical Computer Modelling* 10, no. 8 (1988): 555–61. Reprinted by permission.

conditions, and the like dominate mathematical physics, for example. Such conditions have yielded rich, beautiful, and fruitful mathematical structures, but as we shall see, they are strongly nongeneric. Thus in a certain sense, so too are the material systems which these languages describe.

Material systems whose mathematical images do not satisfy conservation laws or other special conditions of this type are collectively called *open systems*. In particular, biological organisms are open systems in this sense. Indeed, the concept of the open system was first formulated in a biological context more than a half-century ago (von Bertalanffy 1932). The physics of such systems has always posed special problems—problems related precisely to the fact that however hard we try, the nongeneric corpus of contemporary theoretical physics does not apply directly to them. In what follows, we shall exhibit through a few simple examples some aspects of this fact, and its consequences. We shall be specifically concerned with the relation between the closure or isolation of a material system, and the degeneracy of the associated dynamics of the system, mainly in a thermodynamic context. We will then explore some of the interesting consequences of this analysis, both for physics and for biology.

Dynamical Degeneracy

Let us begin with the simplest kind of dynamical system, a linear system in two dimensions:

$$dx/dt = a_{11}x + a_{12}y$$

$$dy/dt = a_{21}x + a_{22}y. \tag{12.1}$$

Generically, the system matrix

$$\mathbf{A} = \begin{pmatrix} a_{11} & a_{12} \\ a_{21} & a_{22} \end{pmatrix} \tag{12.2}$$

is nonsingular. One geometric consequence of this fact is that the two lines

$$a_{11}x + a_{12}y = 0 \tag{12.3a}$$

and

$$a_{21}x + a_{22}y = 0 \tag{12.3b}$$

are distinct and hence intersect at a single point (in this case, the origin). That point of intersection is thus the unique critical point of system 12.1, which we shall call the *steady state* of system 12.1. The stability properties of this steady state are studied in every text on elementary differential equations.

Now let us impose a special condition on system 12.1, which will make it nongeneric. Namely, we will require that the determinant of the system matrix vanishes—i.e., **A** is singular. This means that one of the eigenvalues of **A** must vanish, and entails a number of further consequences.

For example, $|\mathbf{A}| = 0$ means that the two initially distinct lines (12.3a, 12.3b) merge into a single line. Hence every point on this pair of coincident lines is a critical point of system 12.1. We may take as the equation of this line the relation

$$a_{11}x + a_{12}y = 0, \tag{12.4}$$

which, for reasons soon to become apparent, we will call the *equation of state* of the system. We will call any point on this line an *equilibrium point* of the system; the equation of state 12.4 is thus the locus of the system equilibria.

A second consequence of $|\mathbf{A}| = 0$ is that there exists a single number r such that

$$a_{11} = ra_{21}, \ a_{12} = ra_{22}.$$

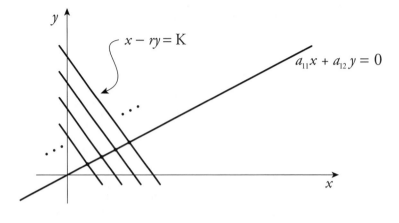

FIGURE 12.1

Hence we can write

$$dx/dt - r dy/dt = 0. \tag{12.5}$$

This means that our dynamical equations in system 1 are no longer independent. Relation 12.5 can be immediately integrated, to give

$$x - ry = K, \tag{12.6}$$

where K is a constant of integration. We will call relation 12.6 a *closure condition*.

In effect, the singularity of **A** has caused our initially two-dimensional system 12.1 to become a one-parameter family of one-dimensional systems, parameterized by the constant of integration, K, in relation 12.6. The phase portrait of this system is shown in figure 12.1.

One can see immediately that this phase portrait is completely different from those arising in the generic case. Instead of a single point attractor, we have a whole line of critical points (equilibria) given by relation 12.4. The trajectories of the system are a one-parameter family of parallel lines, each intersecting the line of critical points at a single point. Specifying initial conditions (x_0, y_0) determines one such line, through the relation

$$x_0 - y_0 = K,$$

and hence a unique equilibrium point to which the system will tend (if the nonzero eigenvalue of **A**, necessarily real, is negative).

Before proceeding further, let us pause to relate our terminology to physical considerations. Our closure condition 12.6, which is a direct consequence of the singularity of **A**, is most familiar when $r = -1$. Then $x + y = K$, the condition which expresses the physical situation in which neither x nor y is entering or leaving the system—i.e., the system is closed. Conversely, if we mandate that a system must be closed in this sense, we force the system dynamics to be singular.

Likewise, in classical thermodynamics, the "equation of state" describes precisely the locus of equilibria of a closed, isolated thermodynamic system—e.g., a gas. A typical equation of state is the Ideal Gas Law,

$$pV = rT, \tag{12.7}$$

where p, v, and T are the thermodynamic state variables. At constant temperature, this relation defines a corresponding curve (isotherm) of equilibria in the (p, V)-plane. Taking logs in equation 12.7 and using temperature units so that numerically we have $T = 1/r$, gives us precisely a relation of the form 12.4 above. It will be seen that we can do a lot of thermodynamics, even irreversible thermodynamics (i.e., dealing with the approach to equilibrium) in this utterly simple mathematical context.

Now let us turn to the three-dimensional case,

$$\frac{dx_i}{dt} = \sum_{j=1}^{3} a_{ij}x_j, \qquad i = 1, 2, 3, \tag{12.8}$$

and repeat the analysis given above. Once again, we shall consider the consequences of making the system matrix

$$\mathbf{A} = (a_{ij}), \qquad i,j = 1, 2, 3,$$

singular.

As always, the singularity of **A** means that one or more of its eigen-values vanishes. In this three-dimensional case, then, we may have either one or two vanishing eigenvalues; this is associated with the rank of **A**. Specifically, if **A** is of rank two, then only one eigenvalue vanishes; if **A** is of rank one, then two eigenvalues vanish. We thus have a choice of "how singular" **A** is to be; this is in turn associated with "how closed" the corresponding physical system is.

For instance, if **A** is of rank two, then there is a single closure condi-tion relating the three state variables, and the locus of equilibria is a line. Our initial three-dimensional system has become a one-parameter family of two-dimensional systems. The diagram analogous to figure 12.1 for the rank two situation is as shown in figure 12.2. Such a system is in effect closed to two of the variables and open to the third. What happens in each of the phase planes depends on the coefficients of the matrix **A** (i.e., on the two free eigenvalues).

If **A** is of rank one, then the manifold of equilibria is now two-dimensional; our initially three-dimensional system (12.8) has become a two-parameter family of one-dimensional systems. The system is now completely closed to all of the variables. This most degenerate situation

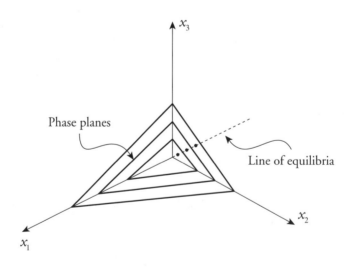

FIGURE 12.2

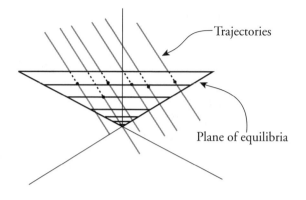

Trajectories

Plane of equilibria

FIGURE 12.3

is shown diagrammatically in figure 12.3. Detailed analysis is left to the reader.

Conversely, if we are mandated a single equation of state, like the Ideal Gas Law (with temperature now in arbitrary units), this simultaneously mandates the rank one situation. That is, the locus of equilibrium points is forced to be two-dimensional; this is as large as it can get without rendering the system completely trivial. This in turn leaves only one dimension free for any dynamics at all.

The same analysis can be repeated for the general linear dynamical system of the form 12.8 (i.e., with any number N of state variables). The salient features of the analysis are

1. Closure of the system is associated with singularity of the system matrix **A**.

2. "How closed" the system is can be measured by the rank of **A**, and hence by the number of vanishing eigenvalues.

3. If **A** is of rank $k < N$, then the original N-dimensional system collapses into an $N-k$ parameter family of k-dimensional systems.

4. If the system is governed by a single equation of state, then it is completely closed; its matrix is of rank one, and its manifold of

equilibria is of dimension $N-1$. This is the maximally degenerate situation.

The same kind of analysis can be extended to more general nonlinear systems, in terms of degeneracies of the Jacobian matrix.

We can see clearly from this analysis how the physical systems which have received the deepest study, and which are the richest mathematically, are also the most degenerate and nongeneric. The same is true for other kinds of physical systems—e.g., the conservative systems of classical mechanics. We shall present a more detailed analysis of these situations, in the context of an extended treatment of the relation between closure and degeneracy, elsewhere. For the moment, we will simply point out some of the profound consequences of this situation.

Discussion

The distinction between closed and open systems is not merely a technical matter. It touches on some of the most basic issues of system epistemology—the distinction between system and environment, between inside and outside. Consequently, the characterization of system closure by intrinsic aspects of dynamical degeneracy has a number of implications along these lines, and serves also to clarify some issues which have long been controversial. We shall touch on a few of these in the present section.

Closed Systems versus Autonomous Systems

An autonomous dynamical system is one for which the time variable does not enter explicitly into the dynamical equations. For this reason, such systems are often called *state determined.*

Technically, the distinction between system and environment appears in the fact that the former is explicitly modeled, while the latter is not. Thus, the effect of a changing environment appears in this context as a time dependence of system parameters, and/or the appearance of additive time-dependent forcing terms. Consequently, the dynamical laws governing such systems necessarily contain time explicitly; they are nonautonomous. Hence intuitively a nonautonomous system must be open. The converse, however, is not true; an autonomous system may be, and generally is, wide open. Consider for example the utterly simple and familiar one-dimensional system

$$dx/dt = ax, \qquad (12.9)$$

which describes an autocatalytic growth or decay. This is an autonomous equation. But it blatantly violates mass conservation; if a system is to obey equation 12.9 (as many do) it must somehow be coupled to its environment in a way not explicitly specified by equation 12.9; i.e., it must be open. In other words, there must be "flows" between system and environment to sustain equation 12.9, which are not visible in the equation itself.

In a nutshell, a system which is both open and autonomous, as equation 12.9 is, must have the property that *the flows from environment to system, and from system to environment, are determined by what is inside the system.* This is indicated schematically in figure 12.4. Here the broken arrows reflect the dependence of rates of inflow and outflow to the system by the state of the system itself, independent of the "state" of the environment (which cannot even be specified at this level). Just on accounting grounds, the dynamics of the situation are as follows:

$$dx/dt = \text{(internal sources and sinks)} = \text{(rate of inflow)}$$
$$- \text{(rate of outflow)}.$$

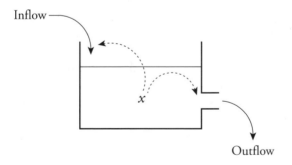

Inflow

x

Outflow

FIGURE 12.4

Since all three terms are functions of internal state x, they are amalgamated into a net rate of change, as in equation 12.9, in which the flows disappear from explicit view. But as we saw in the preceding section, it is precisely on these flows that the genericity of a system like equation 12.9 depends.

It is interesting to observe that an open autonomous system, which is controlling its inflows and outflows through its own internal state, is thereby rendered relatively *independent* of environmental state. That is, it acts as a homeostat. This homeostatic independence from ambient environmental conditions is a ubiquitous characteristic of organisms; it serves in effect to *close* the system from certain environmental conditions, but only at the expense of opening the system to others. These facts have a variety of peculiar implications, not only for an understanding of homeostats, but also for the complementarity phenomenon of *adaptation.* However, since these matters are somewhat removed from the present topic, we shall not pursue them further here.

The Closed System as a Reductionistic Unit

As we have noted above, the corpus of contemporary physics consists of systems which are, in one way or another, closed, isolated, conservative, symmetric, etc. In this light, the "reduction" of biology to physics means precisely the characterization of organic phenomena in such terms. In particular, it means the expression and characterization of the dynamically generic properties of open systems in terms of inherently nongeneric physical mechanisms. Much of the difficulty we experience at the physics-biology interface arises precisely here.

In purely mathematical terms, the situation is analogous to attempting to characterize the properties of generic (e.g., invertible) matrices in terms of rank one matrices, instead of the other way around. The difficulty, which is clearly visible in this mathematical context, is that the rank one situation is so degenerate that essentially anything can happen when such a matrix is perturbed. In physical terms, this means that the properties of a system which is open cannot be inferred from its closed counterpart alone; we must also know specifically *how* the closed system has been opened. Stated another way, there are no "physical laws" available which govern the transition from closed to open, analogous to those which govern systems which are already closed. The closest we come are various results in bifurcation theory, such as the Thom classification theorem (Thom 1976). But these results are not easy to interpret in physical terms (Rosen 1977).

At issue here is the profound question of how to create a physics of open systems—specifically, a physics which will be powerful enough to encompass the material basis of organic phenomena. The prevailing approach, which has developed historically over a period of centuries, is to take the closed, isolated, conservative system as primary, and attempt by one means or another to open it up (Nicolis and Prigogine 1977). The considerations sketched above indicate that this is at best the hard way to go about it; as we have indicated, the closed systems are so degenerate that essentially anything can happen. The alternate strategy is to acknowledge the primacy of open systems, and forget about closed systems in this context altogether. We have indicated elsewhere several possible approaches along these lines (Rosen 1985b, c).

Some Consequences of Dynamical Degeneracy

The essence of a degenerate situation is that it renders otherwise distinct circumstances indistinguishable. A singular matrix, for example, is many-one; it maps distinct vectors in its domain onto a common vector in its range. The more singular a matrix is (i.e., the lower its rank), the larger is the set of distinct vectors which get mapped onto a common image.

Thus a rank one matrix, regardless of its size, can "see" only one dimension, or coordinate, of the vectors in its domain. Thus in a rank one linear dynamical system like equation 12.8 above, the tangent vector to a state will be assigned according to the value of a single coordinate, regardless of the values assigned to the other coordinates. In other words, as we close a system, essential features of its own state become progressively "invisible" to the system.

A closed system moving toward one of its equilibria is generally said to be *disordering.* Many attempts to apply classical thermodynamics to biology seek precisely to analogize such disordering with biological phenomena. Prime among these is *senescence,* which, from the viewpoint of classical physics, is almost the only biological phenomenon which goes "the right way." A common thread running tacitly through much of the theoretical literature on senescence (particularly manifested in the so-called error theories) is precisely that as a biological system ages, it progressively closes itself from its environment. As it does so, then, its behavior becomes more and more independent of its state, until it eventually ends up at some thermodynamic state of equilibrium.

The picture we have developed above offers an alternative to this view. Namely, we can consider what happens to a partially open system when it becomes more open (e.g., the rank of its matrix increases) as well as when it becomes more closed. In the present context, this means that the system "sees" more features of its own states, rather than fewer; the tangent vector to a state depends on more dimensions of the state, and the dynamical behaviors of the system are thereby enlarged. Paradoxically, this very fact can result in a diminishing of adaptive capabilities, as we have shown elsewhere (see chapter 16). Thus, the "physics" of senescence may not be as straightforward as has sometimes been thought.

The mechanisms through which a physical system can modify its own state of closure are themselves of considerable interest for biology. In general, we can see that closing a system (i.e., increasing its dynamical degeneracy) involves the introduction of relationships between system parameters which were initially independent. In inorganic systems, system parameters are rigidly independent of internal state; closure relationships must be physically imposed from outside. To *autonomously* modify a state of closure means making our system parameters state dependent; this in turn requires the introduction of a new family of state-independent parameters, and hence introducing a new hierarchical level. We may perhaps speculate that it is at this level that the biological genome manifests itself most directly (Rosen 1984a).

CHAPTER 13

ᜦ

Some Random Thoughts About Chaos and Some Chaotic Thoughts About Randomness

The class of behaviors collectively called *chaos* represents a milestone for mathematics in general and for the theory of dynamical systems in particular. However, the *significance* of chaos, as a tool for the study of material phenomena, is by no means as clear-cut. Particularly is this true in biology, where it is increasingly argued (e.g., see West 1990) that "healthy" biological behavior is not a manifestation of homeostasis, but rather of chaos—indeed, that homeostasis is pathological. Similar arguments rage in the physical sciences, as for instance in the role of chaos as a paradigm for turbulence, and even for quantum phenomena. In this chapter, I will concentrate on biology, and especially on the impact of chaos on the rationale and philosophy of dynamical modeling of biological processes.

At root, as we shall see, the debate over chaos in biology represents a clash between two distinct traditions in dynamical modeling in general. One of them is based on the concept of *autonomy,* in which every force acting on a system is a function of its phase, or state. In such a case, time does not enter explicitly into the equations of motion; rate of change of state is a function of state alone. *Autonomous* must not be confused with *closed.* On the other hand, an equally strong tradition admits time-dependent forces, independent of state or phase; these are not only allowed but are the central object of study. They are regarded as *inputs,* or forcings, or controls; and the trick here is to express corresponding system behaviors, or *outputs,* as explicit functions of them.

The concept of chaos has enormously extended our understanding of how complicated autonomous behaviors can be, without any external forcings, or state-independent, time-dependent terms entering into the equations of motion. At root, the claim of "chaotics" in this regard is that there is enough complication in autonomous dynamics alone to comprehend basic biological behaviors. The contrary claim is that these same biological behaviors are, in some sense, a superposition, consisting of a much simpler (homeostatic) autonomous dynamics, together with effects of time-independent, nonautonomous, external forcing terms, which could very well be taken as completely *random.*

I shall examine some of the historical roots of the concepts involved, and investigate some of their epistemological correlates, which are profound indeed. Finally, I shall consider the role of "data" or experiment to distinguish which, if either, is correct—in a sense, which are the biological Ptolemaics and which the Copernicans.

Some History

What is now called dynamical system theory grew, in large part, out of the work of Henri Poincaré on the *N*-body problem. He was interested in the stability of a system of *N* gravitating particles initially in general position.

The strategy for approaching such problems in those days was to (1) write down the equations of motion of the system, from Newton's laws; this was (in principle) straightforward; (2) solve the equations, to obtain solutions expressing phases as explicit functions of time; and (3) take the limit $t \to \infty$.

The tricky part in executing this strategy was (2). It had long been known that, *generically,* the equations of motion of such an *N*-body system cannot be analytically solved in *closed form;* they cannot be tidily written in a syntax based on "elementary functions." Rather, they take the form of infinite series of such functions. True, we can extend the series through as many terms as we like, and hence predict the behavior of the system up to any specific *finite* time, but they do not allow us to pass to the limit $t \to \infty$, and hence cannot answer the asymptotic stability question that Poincaré was addressing.

To deal with such problems, Poincaré reasoned, we need a different strategy, one not based on traditional solutions at all. These solutions, he argued, pertain in any case only to finite time, whereas *most* of the time (in the asymptotic realms in which he was interested) the finite-time behavior was quite irrelevant. Moreover, the only reason we want the finite-time solution at all is merely to pass to the asymptotic limit. So he proposed a new strategy to get at the asymptotics directly, without ever having to detour through the finite-time solutions.

It is ironic to note that Poincaré's motivation for generating his new strategy was precisely the failure of predictability for large times, even in these closed, conservative systems. It arises here from the transcendental nature of the finite-time solutions, no matter how exactly we specify initial conditions. The lack of predictability, today considered a hallmark of chaos, is in fact nothing new.

Since, as Poincaré realized, any trajectory will spend "almost all" its time near its asymptotic limit, we need only to characterize the limits themselves, what he called the ω-limiting sets of trajectories, including what we today call the *attractors,* directly from the equations of motion. Since we do not need explicit solutions of the equations in this procedure, it was christened *qualitative.* Poincaré himself solved the problem in two (phase) dimensions, where he showed that the only possible limiting sets are point-attractors and (limit) cycles. Of course, two dimensions are too few to be of interest for N-body problems, and Poincaré himself was well aware that more complicated things could happen in higher dimensions. But it was a start and indeed a most auspicious one.

Poincaré's arguments are based on a partition of system behavior into two parts: a *transient,* finite-time part, and a *"steady-state,"* asymptotic part. The transient behaviors, expressed by traditional finite-time solutions of the equations of motion, pertain to what is happening in the *basins* of attractors; the steady-state part pertains to the attractors themselves, or more generally, to the ω-limit sets of trajectories. It was Poincaré's idea to dispense entirely with transient behavior and concentrate on steady-state.

Exactly the same partition into transient and steady-state underlies the study of nonautonomous systems and how they respond to time-dependent external forcings. In that situation, it is assumed that the system being forced is *already* at its autonomous steady-state. Thus if we simply subtract this autonomous, steady-state behavior from what the system is actually doing in responding to a time-dependent input or forcing, we get precisely how the system is affected by the forcing alone—the corresponding *output.* The simplest situation is where the autonomous behavior has "damped to zero"; in such a case, the steady-state response of the system to a time-dependent input depends on that input alone; it tells how the output "tracks" its input; it expresses the system "transfer function."

It is absolutely essential in this picture that the time-dependent inputs, or forcings, not change autonomous *dynamics* in any way. It must neither create nor violate the system *constraints;* more generally, it must not change any of the *parameters* that govern both transient and asymptotic behaviors in the autonomous system. Stated another way, any allowable external, time-dependent forcing may tamper only with the initial conditions (initial phase or state) of the autonomous system to which it is applied.

This restriction is not often mentioned explicitly, but it is absolutely basic. Unless we impose it, the very identity of the system will depend on the environment with which it is interacting. That is, the identity of the system in vacuo will be different from what it is in a specific context that is forcing it; that identity has become, in a sense, *context dependent.* The autonomous system itself, instead of providing the basis for all subsequent analysis, becomes rather an *artifact;* it is literally a different system in isolation from what it is in interaction. Thus it is that parametric control is a very tricky business, but that is another story. Note that the only inputs or time-dependent forcings compatible with the context-independence of identity of an autonomous system are those I have envisioned here, and that this is an exceedingly strong condition to impose on system *environments.*

The N-body problem, Poincaré's point of departure, is based on these assumptions. If we put these N bodies into a universe otherwise empty, then we get the autonomous situation, with no external forcings at all. If there are other gravitating particles in that universe, with which our original N bodies can interact, then we can follow one of two strategies. One is to lump the effects of these particles into a (small) *random,* time-dependent fluctuation or forcing term—a perturbation, *noise.* The other is to subsume these other particles into the original system— that is, to pass from the original N-body problem to a new $(N+k)$-body problem, and recover autonomy in the larger system. If we do this, we in effect turn what was noise in the smaller system into an explicit function of phase in the larger system. The two are supposed to be equivalent; but, at best, they are equivalent only if we mandate that a nonempty environment can affect only autonomous system *behavior,* never its autonomous dynamics or identity.

On Genericity

The experience of (classical) physics is that the language of differential equations is the *generic* language in which to describe material reality.

Furthermore, chaos is telling us that what have been dubbed *strange attractors* are, in a precise sense, *generic* in autonomous dynamical systems. Or, what is the same thing, that trajectories consisting of single points (equilibria, or steady states) and closed curves (limit cycles) are rare among the limiting behaviors manifested in dynamical systems.

Putting the two together, chaotics is arguing that it will correspondingly be generic for those dynamical systems that are realized by, or provide models for, material phenomena to have strange attractors. Or, even more generally, that most trajectories in such systems will not be simple curves in phase (state) space, but will be more like fractals (West 1990). This is indeed true in conservative systems (themselves already highly nongeneric), where there are no attractors at all, but in which most trajectories are *ergodic*—dense in their energy surface.

On the other hand, it is equally possible to argue that autonomy itself is a rare, nongeneric property in the real world of material phenomena. *Most* systems, in a sense, are coupled to, or interacting with, definite environments. The environments are not themselves included in our models of the *system,* except insofar as they impact on system behavior (i.e., its dynamics). And *generically,* this impact will be expressed by state-independent, time-dependent forcing terms, tacked on to the inviolate autonomous dynamics.

If this is so, then the genericity of strangeness in autonomous systems becomes much less important as a modeling strategy. We can very well admit that autonomous material phenomena *do* indeed give rise to very special, nongeneric dynamical models, as long as we put the emphasis on interactions with particular environments. That is, as long as we shift what is generic from autonomous behavior to interactions.

Indeed, genericity itself is a two-edged sword. On the one hand, as the name itself implies, what is generic is unencumbered by the *special* or the specific, and independent of the particularities that set one dynamical system apart from another. On the other hand, the more specifics we have, the more conditions we impose, the more theorems we can prove. That is precisely why an objectively very special class of dynamical systems, the conservative ones, can be (and have been) studied in such detail, both in themselves and as models for material nature.

Indeed, the very success with excessively nongeneric systems such as the conservative ones argues a contrary strategy—namely, to assume the simplest *autonomous* dynamics, in some sense the most nongeneric we can get away with, and put everything else into environmental interaction or forcing terms. Since, by the *definition* of the forcings that are allowed, the nongeneric *autonomous* dynamics will not be changed by them, and hence its inherent nongenericity will not be exposed by them.

But only very nongeneric *environments* will satisfy this kind of condition. In a sense, then, the real question is whether there is more genericity in autonomous strangeness, or in nonstrange autonomous dynamics supplemented by allowable temporal forcings.

"Free Information"

The concept of "free information" was introduced by Sir Arthur Eddington, in a curious and much underappreciated book called *Fundamental Theory,* published posthumously in 1946. Here is how Eddington introduces the concept:

> In theoretical investigations we do not put ourselves quite in the position of an observer confronted with an object of which he has no previous knowledge. The theorist is considering, let us say, an electron with coordinates x, y, z. He recognizes that knowledge of x, y, z could only be obtained by observational measurements performed on the electron, and that these measurements will create a conjugate uncertainty in the corresponding momenta. But knowledge of the mass m and charge e is on a different footing. Their values are taken from a list of physical constants. It is true that these values of m and e rest on observation, but not on observation of the object at x, y, z. As applied to that object they are *free information*—not to be paid for by a reciprocal uncertainty of the variates conjugate to them.   (p. 19)

An observer confronted with an unknown particle does not start with knowledge that it is an electron. If he measures the mass and charge he may be able to infer that it is almost certainly an electron; but meanwhile he has by his measurements created reciprocal uncertainty in the conjugate variates. A pure observer cannot obtain free information. On the other hand, an actual observer is a human being, and therefore an inveterate theorist; and he will probably proceed to steal free information by substituting the accurate tabular values of m and e for his own rougher measurements without paying the enhanced price.   (p. 20)

Notice here the tacit identification of theory with the stealing of such "free information"—information to which a pure observer is not entitled. This is as good a way of characterizing theory as I have ever found—theory in the good old sense of "theory of functions of a complex variable," or "theory of ordinary differential equations," or "theory of numbers," not in the dictionary sense of "a plan or scheme existing in the mind only; a speculative or conjectural view of something." As such, it has to do not so much with *data* but with the extrapolability of data; it pertains to similitude (Rosen 1983).

Our concern here is with the notion of chaos as a source of such "free information" about dynamical processes in material nature, and particularly about biological processes. Specifically, that chaos per se, as a thing in itself, provides insight into nature at large, independent of what specific phenomenon or process we may be looking at. It permits us, for instance, to study turbulence as a thing in itself, independent of what specific fluid is turbulent. A pure observer would rather say that one may study, for example, only turbulent water, or turbulent air, in which *turbulent* is only an adjective. Chaotics would say rather that *turbulence* is a noun, and that any particular fluid is the modifying adjective (see the section On Nouns and Adjectives in chapter 9). In short, an independent concept of turbulence becomes a source of free information about water, or air—information that one does not need to *observe* water, or air, or any other particular fluid, to know.

All this provides another way of stating that we can argue from a generic fluid to a particular one. The argument here is, roughly, that since almost all dynamical systems have strange attractors, therefore a given particular one will. This is precisely the free information about particulars that a generic property offers—information not available to a pure observer. Indeed, the psychological allure of chaotics lies precisely here—the offer of great benefit at little cost.

There is, in fact, a second source of free information in this picture, arising from the employment of asymptotic, limiting behavior (the attractor) to tell us things about finite-time behavior, about what I have called *transient*. Indeed, unless we are already *on* an attractor, or more generally, on an ω-limit set of a trajectory, we are in a transient realm, and not an asymptotic realm at all. On the other hand, since basins' (transient, or finite-time) behaviors are generically much bigger than their attractors, almost all initial conditions in the basin will not lie on the attractor (i.e., on the steady-state, asymptotic regime) at all.

Nevertheless, segments of time-series, pieces of finite-time, transient response, such as electroencephalogram or electrocardiogram traces, are often invoked to bolster a chaotics view, a view that the signal represents generic, autonomous, asymptotic processes rather than simple (i.e., nongeneric) autonomous activities externally forced by noise in a transient domain.

General Discussion

As we have seen, chaotics puts genericity entirely into autonomous dynamics, whereas the more traditional view puts it into the environment, in the form of (random) forcings. These views, different as they are, are argued on similar grounds, as inductions from experience or data, and, in fact, from precisely the same data. This means that such data do not suffice to force acceptance of one picture or the other, nor indeed even to discriminate between them.

Nevertheless, the two pictures are poles apart conceptually. They are most different in terms of the free information we can steal from them about what is really going on. The situation is quite analogous (and indeed, closely related to) the program theories and the error theories that have been put forward to explain senescence (Rosen 1978c)—the former as faithful execution of internal (autonomous) program, the latter as environmentally forced departures from such a program. Again, both are based on the same data and cannot be discriminated on that basis.

The difference between such polar views of what is happening in organisms is far more profound than, say, between the Heisenberg matrix mechanics and the Schrödinger wave equation in quantum mechanics. These cannot be distinguished because they are abstractly identical in mathematical terms—alternate realizations of the same underlying abstract, objective structure. This is not true in the situation we have been discussing; the two pictures are *objectively* different, but we can nevertheless *simulate* or mimic the one by the other. But the two pictures lead to entirely different *models* of dynamic phenomena in biology.

Indeed, the traditional picture (simple, nongeneric dynamics, contaminated by environmental forcing) is much more suited to making and interrogating actual models of biological behavior, just as Hamiltonian, conservative dynamics is in physics. Indeed, I know of very few actual (biological) *models* that can claim chaotic attractors at all (at least in continuous-time dynamics; discrete-time models are very different, as I discuss later). Rather, the invocation of (continuous-time) chaos in specific biological contexts is based on the indirect argument that chaos does arise in the same contexts as the actual models do (e.g., in mass-action systems, as are used to model biochemical kinetics, ecosystem behavior, and population dynamics generally). This would seem to undermine the basis of the free information that we can argue from "almost all" (i.e., from generic) to "all" (i.e., to particular). However, models are typically preselected on the basis of tractability, and, as Poincaré realized, tractability in this sense is already highly nongeneric.

By its very nature, nongeneric dynamics is inherently less *robust,* in the face of truly arbitrary environments. Indeed, nongeneric dynamics can look robust only if we severely restrict the genericity *of the environment* as a source of system fluctuations. Perhaps, indeed, the way to discriminate between "chaotics" and its alternatives lies precisely here—pushing a system into a truly arbitrary niche and seeing what happens. It is, of course, equally possible that both pictures are wrong (Rosen 1979), that dynamics itself is already too nongeneric.

Finally, *discrete chaos* is usually viewed simply in terms of a sampling of the behavior of a continuous-time dynamics at a set of discrete instants. But the relation between continuous-time and discrete-time dynamics can be much more subtle than that, as Poincaré pointed out long ago. Based on his notion of *transversality,* the crossing of a continuous-time trajectory through a lower-dimensional subspace, so that the time-index in the discrete system becomes ordinal rather than cardinal, there is not even any continuous time of which the discrete time is a sample. The relation between the two systems is usually called a *suspension* of a discrete system in a continuous one; conversely, starting from a discrete system and trying to find a corresponding continuous dynamics generically requires *more variables,* a higher-dimensional state space, and vastly different equations of motion. Thus although there are many parallels between continuous and discrete chaos, they are in many ways quite different animals. Thus it is that, for instance, a continuous-time version of the logistic equation is a very docile beast, while its discrete-time counterpart can be very chaotic indeed. On the other hand, once we get into a discrete-time situation, chaos will tend to appear generically in actual biological models.

It is hoped that this chapter will help clarify the claims and counter-claims provoked by the notions of chaos as a new scientific paradigm.

Part IV

∾

SIMILARITY AND DISSIMILARITY
IN BIOLOGY

T HE CHAPTERS in this part cover a variety of more special topics
and applications, especially to biological form and to morphogenesis. Chapter 14 was prepared originally as a tribute to Richard Bellman, who was perhaps best known as the creator of a powerful optimization technique that he called dynamic programming. This technique
is closely related to the Hamilton-Jacobi equation, which in turn arises
out of the mechano-optical analogy of William Rowan Hamilton, one
of the most profound innovations of nineteenth-century physics. It was
based on the observation that both the paths of mechanical particles in
configuration space, and the paths of light rays in optical media, are
governed by minimum principles (least action in mechanics, least time
in optics). Making these principles correspond allows, as Hamilton
showed, the establishment of a precise "dictionary" between optical and
mechanical phenomena. That is, it provided a way of establishing a
deep relation between these two disparate branches of physics, without
any attempt to reduce either to the other.

The power of this analogy was so great that Hamilton himself could
have, via the Hamilton-Jacobi equation, which describes the spreading
of "wave fronts" (Huyghens' principle in optics), derived the Schrödinger equation, the basis for the form of quantum theory known as
wave mechanics, almost a century before Schrödinger himself did it.
From my very early student days, I had been much impressed by the
unique power of this nonreductionistic approach, and I subsequently
used it to develop my ideas about modeling relations in general, and
their unique roles in science. These ultimately took a category-theoretic
form, whereas what Hamilton was doing was of a functorial character.
Thus the Hamilton analogy provided one of my own early bases for

believing that mathematical homologies via modeling provided a powerful alternative to reductionism.

Moreover, these ideas, in the 1950s and 1960s, provided the mathematical basis for the then-emerging optimal control theory, to which Bellman's dynamic programming belonged (as did, in another way, the roughly equivalent Pontryagin principle, based on paths or rays rather than on wave fronts).

Finally, there was the intimate entangling of ideas regarding optimality with evolution and fitness. Ordinarily, one cannot assess the evolutionary fitness of a phenotype, or of its underlying genotype, by looking at either of them; rather, one must look at progeny. Rashevsky's principle of optimal design provided a way around this peculiar situation, by expressing fitness in terms of design criteria of the actual phenotype by itself. This was a daring departure from the prevailing ideas of the time regarding the nature of fitness, and of how selectional or evolutionary processes themselves were driven.

Optimality also underlay all the canonical form theorems of mathematics. That is, it provided a way to reach into equivalence classes and extract single representatives that also satisfied additional, superimposed optimality criteria. In a nutshell, optimality in this sense *is* selection. It turns out, furthermore, that the converse of this proposition is also essentially true; selection criteria generally can be turned into optimality principles of a particular kind.

Chapter 14 concentrates on the aspects of optimality pursued in a primary way by Bellman, and it combines these with a broad view of optimality as a research strategy. Of particular interest is the circle of ideas associated, in the classical calculus of variations, with the Noether Theorem, which serves to directly tie optimality with ideas of similarity, and with conserved quantities (invariants).

Chapter 15 describes an approach to morphogenesis, an area with which I have long been concerned. It sketches a simple version, with some specific examples, of an integrated approach to morphogenesis in spatial arrays of interacting elements, which I call morphogenetic networks. Optimality ideas pervade almost all of it, but they are not stressed. Rather, the chapter focuses on generalizations of what are usually called modular networks, which range from the inorganic Ising models of physics, and the purely mathematical cellular automata (including neural networks of all kinds), to cellular populations and even ecosystems. What holds all these diverse entities together is a commonality of underlying structure—a mathematical form. The novelty here is in the interpretation of a given set of states, associated with each element in the network, as consisting of distinct classes. This reflects a prior classification of morphogenesis into three basic "morphogenetic mechanisms," initially put forward by John Tyler Bonner (who edited and wrote an introduction for an abridged version of D'Arcy Thompson's *On Growth and Form* in 1961), and which I describe as (1) sorting, (2) differentiation, and (3) differential birth-and-death. A general morphogenetic process, in this framework, involves the three kinds of mechanisms proceeding in parallel, and comparing the results with what each of these mechanisms can accomplish alone. It encompasses, in principle, all kinds of morphogenetic processes, from the folding of a protein molecule to the distribution of species into niches on a landscape. There is a potentially large research program tacit in these morphogenetic networks; certain very special cases have been pursued, but it has never been taken up in its full generality.

Chapter 16 was motivated, in part, by biological phenomena of senescence and my approach to it under the rubric of anticipatory systems. In anticipatory systems, as I have defined them, the present change of state depends on a future state, which is predicted from present circumstances on the basis of some model. Anticipatory, model-based behavior provided one basis for what I later called complexity, and which I defined in *Life Itself* on the basis of noncomputability or nonformalizability of models of a system of this kind. Such systems generally possess a number of interesting properties, as a consequence of their necessary complexity. One is that they possess intrinsic semantic aspects, associated with a loop linking the present value of some quantity with a future value of another (generally different) quantity via the predictive model, and then that predicted value with the present rate of change of others (the anticipatory control). The behavior of such anticipatory systems is generally quite invulnerable to simple-minded reductionistic fractionations. Such systems can also senesce, in a mode I have called global system failure, which is different from a local failure, in which some particular component malfunctions (these are characteristic of simple systems only). Finally, there are what I call the *side effects of simple interventions* in such systems, arising from the fact that models tend to be more closed, in a physical sense, than the systems being modeled. This is what chapter 16 is about, albeit in a rather limited context. Such side effects, which measure the difference between what we think we are doing (on the basis of simple models), and what we are really doing, are of crucial significance to the way we think of systems, but they have never received the study they should—the conceptual basis for doing so is not present.

In chapter 17, I have tried to indicate something of the ontics, rather than the epistemics, involved in producing a synthetic organism. The chapter needs to be supplemented (e.g., with a deeper discussion of what *existence* means), but I hope enough is presented to indicate some of the flavor of the ontic arguments involved.

CHAPTER 14

ॐ

Optimality in Biology and Medicine

Optimality is the study of superlatives. Nowadays, optimality plays two distinct but interrelated roles in science: (1) in pure science, its impact is primarily analytic or explanatory; we use it to characterize the way in which a natural process *does* occur, out of all the ways it *could* occur; and (2) in applied science (technology or engineering in the broadest sense), we use it to decide how we *should* do something, out of all the possible ways in which we *could* do it (here the emphasis is primarily on design or synthesis).

Like most good scientific ideas, optimality originated with the Greeks. In Greek thought there was much concern with perfection—with what was "best" or most perfect in some abstract sense. Because *best* is a loaded word, presupposing an extrascientific system of preassigned values, the employment of optimality principles in science has to this day not been completely disentangled from metaphysical, theological, and even ethical associations. To give one famous example, no less a person than Leibniz attempted to characterize our entire universe as "the *best* of all possible worlds," mainly because he thought that our world *maximizes* something he believed was good ("compossibility," the number of things or processes which can coexist).

Purely formally, any optimality problem consists of the following three data (Rosen 1967).

1. *universe of discourse U*, which consists of all the potential or possible solutions to a given problem;
2. with each element $u \in U$, an associated number $\rho(u)$, which we shall call the *cost* of u. The mapping $\rho : U \rightarrow \mathbb{R}$ so defined will be called the *cost function* of the optimality problem. We shall

use this terminology even though we may be talking about a negative cost—i.e., a benefit or a profit;

3. a means of determining the element or elements $u^* \in U$ for which the cost is least (or, alternatively, for which the profit is greatest).

In the abstract, the superimposition of a cost function on a universe of discourse is simply a way of picking specific elements out of a class of otherwise equivalent or indistinguishable things. Indeed, most of the "canonical form" theorems of mathematics (e.g., the Jordan Canonical Form of linear algebra) can be reformulated as optimality problems; the role of these theorems in mathematics is to pick representatives out of equivalence classes (i.e., to *reify* the classes) in a systematic way.

The character of an optimality problem depends crucially, and extremely sensitively, on the choice of the universe of discourse U, and of course on the specification of the cost function. In pure science, these are chosen primarily on abstract theoretical grounds—in applied science, on practical ones. We shall discuss these in more detail in the subsequent sections. The actual determination of the solutions of a given optimality problem, however, belongs to mathematics. We shall discuss some of the basic mathematical features of these problems now.

The discussion which follows will concentrate on the more conceptual questions of extremizing cost functions, especially on the relation between extremization and stationarity, and the intuitive significance of the familiar necessary conditions for stationarity. These will be clothed with material substance in subsequent sections. We shall not consider, except most briefly in passing, the construction of explicit algorithms for optimization. These matters cannot be described better than was done many times by Richard Bellman (1957, 1961; Bellman and Dreyfuss 1962) under the rubric of *dynamic programming;* by a host of Russian authors (Gelfand and Fomin 1962; Pontryagin et al. 1961) using what has come to be called Pontryagin's *Maximum Principle;* and available nowadays in a plethora of textbooks. We thus gladly defer to these treatments, and restrict ourselves to those conceptual matters which must precede application of these methods.

In principle, the problem of maximizing or minimizing any real-valued function is trivial, because the topology of the real numbers is generated by a relation of total ordering. Thus if the image $\text{Im}(\rho)$ of a mapping $\rho: U \to \mathbb{R}$ is bounded above, then there is a least upper bound r^*. If Im(ρ) is closed, then this bound is actually attained, and the set $\{\rho^{-1}(r^*)\} \subset U$ gives us the solutions which maximize cost. If Im(ρ) is not closed, then we cannot actually attain the value r^* on U, but we can come as close to it as we please. If Im(ρ) is not bounded above, then there are no solutions; we can find elements $u \in U$ whose costs are arbitrarily great. Likewise, if Im(ρ) is bounded below, there is a greatest lower bound r_*; whether it is attained or not depends as before on the closure of Im(ρ).

In practice, however, things are not so simple. Typically, the cost function ρ is given as a rule which allows us only to compute its value $\rho(u)$ when an element $u \in U$ is given. To determine from this local information whether Im(ρ) is bounded or not, and even more, to compute the bounds, confronts us in general with serious foundational or computability questions. So we usually try to make life simpler for ourselves by imposing further conditions on U and on ρ. For instance, we may require that U be some kind of a manifold; that ρ is sufficiently smooth or differentiable; and that we may replace the bounds r^* and r_*, in which we are really interested, by an associated notion of *stationarity*, which is not quite equivalent to it. This indeed is the framework in which traditional approaches to solving optimality problems were developed.

If we assume further that U is actually a *finite-dimensional* manifold, then we know from elementary real analysis that the elements $u_0 \in U$ for which

$$\text{grad } \rho(u_0) = 0 \qquad (14.1)$$

render ρ *stationary*. This means intuitively that a first-order perturbation $u_0 \to u_0 + du$ will produce only a higher order change in cost, i.e., one which is *negligible* compared to du. At least some of these stationary points u_0 are also *local extrema* (the classical maxima and minima); thus, if u_0 is locally a maximum, there is some neighborhood U_0 about it such that $\rho(u_0) \geq \rho(u)$ for all $u \in U_0$, and correspondingly for (local) minima. Thus, the classical approach to solving an optimality problem in these circumstances is to solve equation 14.1 for the stationary points u_0, and treat this set of stationary points as a new universe of discourse. If ρ is sufficiently smooth and well behaved, there will only be a finite (hopefully small) number of stationary points, and the optima we seek will be among them.

We can actually do better than this. Namely, we can in principle use the cost function ρ to generate a dynamics on U, for which the stationary points are automatically the steady states. Specifically, we can put

$$du/dt \ = \ \text{grad } \rho(u). \tag{14.2}$$

By doing this, we are essentially forcing the cost function to be a Lyapunov function; indeed, we are taking the "simplest" dynamics for which this is true. Most of the algorithms for actually finding extrema are based essentially on this. In the dynamics 14.2, the local maxima will be stable and the local minima unstable, while the other stationary points will be semistable. If we put

$$du/dt \ = \ -\text{grad } \rho(u), \tag{14.3}$$

then it is the minima which will be stable, and the maxima unstable.

If U is not finite-dimensional, then all we can retain of the above is the notion of stationarity. The most typical situation is one in which our universe of discourse U is a space of functions

$$\{f : X \ \rightarrow \ Y\}.$$

For instance, X may be the closed unit interval [0, 1], Y some manifold, and U the space of all functions f for which $f(0) = y_0$, $f(1) = y_1$, where y_0, y_1 are fixed elements of Y. That is, U can be thought of in this case as a set of curves in Y with fixed endpoints.

A real-valued map from such a function space is usually called a *functional*. A typical way of associating a number with a functional in U is through some kind of integration process—intuitively, either an averaging, or a measure of arc length in some suitable metric. Thus, we can generate many such functionals, as follows: Suppose that we are given some real-valued function of the form

$$F \ = \ F(x, \ p_1, \ldots, p_k),$$

where $x \varepsilon X$, and $p_1, \ldots, p_k$ are simply arbitrary independent variables. If we now take a specific function

$$f : X \rightarrow Y$$

in U, we can put

$$p_1 = f(x), \; p_2 = f'(x), \ldots, p_k = f^{(k-1)}(x).$$

Thus the curve $f \in U$, defined originally in Y, can be used to generate another curve in the space of arguments of the given function F. Let us then define

$$\rho(f) = \int F[x, f(x), \ldots, f^{(k-1)}(x)] dx \qquad (14.4)$$

and treat this as a cost function on U. Intuitively, this cost function evaluates properties of f which depend on the values of f and its first $k - 1$ derivatives.

With this kind of functional, the notion of stationarity can still be retained, as we noted earlier. For instance, if we define a "variation" of $f \in U$ to be a function of the form

$$(f + \delta f)(x) = f(x) + \varepsilon g(x)$$

(where ε is a "small" number, and g is any function such that $f + \varepsilon g \in U$) and define the corresponding change (variation) of cost to be

$$\delta\rho = \rho(f + \delta f) - \rho(f), \qquad (14.5)$$

we can ask for those functions f_0 in U for which this variation in cost is stationary:

$$\delta\rho(f) = 0. \qquad (14.6)$$

up to first-order magnitudes. This, of course, is the standard format for the classical calculus of variations. It should be noted that all this apparatus reduces to the finite-dimensional (in fact, one-dimensional) case when we consider only functions f such that

$$f = \text{constant}.$$

As is well known. the solutions of equation 14.6 must satisfy an Euler-Lagrange differential equation. Essentially, this is an expression of functional dependence of the kth derivative $f^k(x)$ on the preceding $k-1$ derivatives

$$f(x),\ f'(x),\ldots,f^{(k-1)}(x),$$

a relation which holds at every point $x \in X$. In particular, if $k = 2$, the Euler-Lagrange condition for stationarity is a second-order differential equation. The solutions of this equation are thus precisely the stationary curves of the cost functional equation 14.4, among which we may hope to look for the actual solutions to our optimality problem.

One peculiarity of the Euler-Lagrange-type conditions should be noted here. Namely, it is extremely difficult to get *first-order* differential equations out of them. One might think that putting $k = 1$ in equation 14.4 above would yield such first-order equations. But if one tries to do so, we find that the term in the Euler-Lagrange conditions that yields *all* the temporal derivatives vanishes identically. This fact alone indicates that there is something peculiar about the systems of first-order differential equations which are universally used as models of kinetic phenomena in chemistry, biology, and elsewhere, in that they seem immune to generation by optimality principles. We shall take up this point again in the next sections.

Optimality in Physics

Probably the earliest effective invocation of optimality principles in physics was the Principle of Least Time of Fermat, dating from 1650. The problem was to determine the actual path a ray of light would take between two points x_0, x_1 of an arbitrary (i.e., inhomogeneous and anisotropic) optical medium. In this case, the universe of discourse U is the set of all paths in the medium between the two points (i.e., the set of smooth maps *f* from the unit interval to the region of space occupied by the medium, such that $f(0) = x_0$, $f(1) = x_1$). As cost function, Fermat suggested the time of transit $\rho(f)$ of the light ray along the path *f*. Since the medium is presumed inhomogeneous (specifically, varying in refractive index from point to point) and anisotropic (varying from direction to direction), the cost function in this case is an integral whose integrand involves both position and velocity:

$$\rho(f) = \int F[x, f(x), f'(x)]dt.$$

Fermat claimed that the paths which extremize this cost function (specifically, which minimize it) are precisely those paths which a light ray will actually follow. This was a most daring proposition, but it was bolstered by the fact that, using it, one could easily account for all the known facts of geometric optics (e.g., the "laws" of reflection, refraction, etc.).

In 1686 Newton published his "Principia," which electrified and revolutionized the scientific world. Newton himself apparently did not think in terms of optimality, although both his strong theological bent and his interest in variational problems are well known. However, around 1700, John Bernoulli was already analogizing the motion of a particle in a field of force with the path of a light ray in an optical medium. Some decades later, we find the speculations of Maupertuis, who hypothesized that every natural process takes place in such a way that a mechanical quantity, which he called "action," was minimized thereby. Similar considerations had occurred to Euler, and played an important part in Euler's pioneering work on the Calculus of Variations. All of this was vastly extended by Lagrange, whose monumental book on mechanics appeared in 1788. These ideas culminated in the first part of the nineteenth century with the work of Hamilton and Jacobi. It is worth briefly reviewing some of the basic ideas, as they have not only been basic to theoretical physics, but also provided the model for more recent work on optimal control.

The key to the relation between optimality and mechanics lies in the circle of ideas surrounding Newton's Second Law. Newton realized that the *configuration* of a system of structureless particles (i.e., the coordinates of the constituent particles at a given instant) do not in general determine the *velocities* of those particles—these can be freely chosen to have any values whatever. One might think that the same would also be true of the accelerations, and indeed of all higher temporal derivatives of configuration. But in an incredible masterstroke, Newton said no; if we know where the particles are at an instant, and how fast they are going, the accelerations are completely determined. Specifically, he said that the rest of the world was exerting *forces* on our particles, and that these forces could be measured by looking at our system alone, by seeing how the particles in it are accelerating as a function of where they are (configuration) and how fast they are going (velocity) at any instant. Thus, Newton posited a recursive relation

$$d^2x/dt^2 = F(t, x, dx/dt) \qquad (14.7)$$

specifying the second temporal derivative of configuration (acceleration) in terms of the first derivative (velocity), the 0th derivative (configuration itself), and perhaps of time explicitly.

In modern language, we would write the second-order relation 14.7 as a pair of first-order relations

$$dx/dt = v$$
$$dv/dt = F(t, x, v), \qquad (14.8)$$

i.e., as a dynamical system. If F is independent of t (i.e., is *autonomous*), this determines the motions of the system of particles as a field of trajectories in *phase space,* which roughly speaking is the space of all position-velocity pairs. The projections of these trajectories on the space of configurations define the possible *paths* of the system in configuration space. It is this last situation which Bernoulli, Euler, Maupertuis, and later Hamilton analogized with propagation of light through optical media.

Now the relation 14.7, relating a second derivative recursively to the lower derivatives, is something we saw in the previous section. Namely, such a relation also characterizes curves which are stationary for some cost functional of the form of equation 14.4, when we put $k = 2$. Thus, from this perspective, it is natural to look for a cost functional whose stationary curves are also the paths of a Newtonian system moving in a given field of force.

If (and *only* if) the system is conservative, this can easily be done. Just to fix ideas, let us consider the simplest such system, the one-dimensional harmonic oscillator with unit mass. Its equations of motion can be written as

$$dx/dt = v$$
$$dv/dt = -kx = \text{force.} \tag{14.9}$$

In this case, the phase space is two-dimensional, and the trajectories are characterized by the curves

$$v^2/2 + kx^2/2 = \text{constant.} \tag{14.10}$$

These are closed curves (ellipses) in the phase space, corresponding to periodic solutions. The relation 14.10 expresses conservation of energy; the first term on the left, depending only on velocity, is *kinetic energy,* while the second term, depending only on configuration, is *potential energy.*

Let us write the relation 14.10 in the form

$$T(v) + U(x) = H(x, v). \tag{14.11}$$

The function on the right, which is constant on the trajectories, is the Hamiltonian, or total energy. We can verify immediately that it has the magic property that *its gradients in the phase space determine the temporal derivatives of the phase variables;* specifically, we have

$$dx/dt = \partial H/\partial v,$$
$$dv/dt = -\partial H/\partial x. \tag{14.12}$$

These relations, which are perfectly general, comprise the *Hamiltonian form* of the dynamical equations 14.9. The reader should note the similarity of these equations with 14.2 and 14.3 above; this similarity is not accidental, as we shall see.

Let us go a little further at manipulating these relations. If we differentiate the first of them and put it equal to the second, we have

$$d/dt(\partial H/\partial v) + \partial H/\partial x = 0 \tag{14.13}$$

We can get this into a more perspicuous form by introducing in place of the Hamiltonian $H(x, v)$, which is the sum of kinetic and potential energy, a new function

$$L(x, v) = T(v) = T(v) - U(x), \qquad (14.14)$$

which is the excess of kinetic over potential. It is readily verified that 14.13 becomes

$$d/dt(\partial L/\partial v) - \partial L/\partial x = 0, \qquad (14.15)$$

which just happens to be precisely the condition of stationarity for the cost functional

$$\rho[x(t)] = \int L dt \qquad (14.16)$$

defined on the space of paths of the system 14.9. In fact, the relation 14.15 is a second-order equation (seen explicitly by expanding the total time derivative on the left-hand side) which is completely equivalent to the original equations of motion, and to the Hamiltonian form 14.12. These are the Euler-Lagrange forms of the equations of motion; the function $L(x, v)$ introduced by 14.14 is the *Lagrangian*. The cost functional 14.16 itself is defined as the *action*. This is the origin of the celebrated *Principle of Least Action* for characterizing the paths of (*conservative!*) systems of Newtonian particles in fields of force.

This circle of ideas is one of the most powerful in all of theoretical science. Let us give a few indications of just how powerful it is. First, by appropriately analogizing the Principle of Least Action with Fermat's optical Principle of Least Time, Hamilton was able to establish a complete "dictionary" between the two areas of optics and mechanics. This "mechano-optical analogy" related the two disparate branches of physics in a new, *nonreductionistic* way. This was in contradistinction to other scientists (e.g., Maxwell) who tried (unsuccessfully) to *reduce* optics to mechanics, or to Einstein's later search for the "unified field." Since Hamilton's time, the establishment of such analogies, in terms of optimality principles, has been a major theoretical tool in the investigation of new classes of physical phenomena.

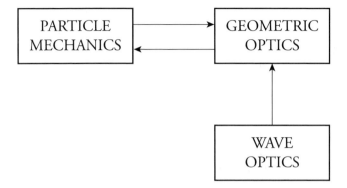

FIGURE 14.1

Hamilton went still further. Guided by the fact that geometric optics, characterized by Fermat's Principle, is a limiting case of a more microscopic theory of *wave* optics, he was able to write down the partial differential equation expressing the propagation of surfaces of constant action in the configuration spaces of conservative mechanical systems (the Hamilton-Jacobi equation). He thus established a relation between particle mechanics, geometric optics, and wave optics which is schematized in figure 14.1. If he had had the least reason for doing so, he could have surmised that there might be a more microscopic theory, sitting under particle mechanics in the same way that wave optics sits under geometric optics. The contents of that hypothetical box might be called "wave mechanics"; in that case, the Hamilton-Jacobi equation would transmute into precisely what is now called the Schrödinger equation (and indeed, this is essentially the way in which Schrödinger derived his equation, nearly a century later). Such is the power of the mechano-optical analogy, based on relating disparate classes of material systems through homologous action principles (i.e., through homologies of mathematical description) rather than through any kind of reductionism.

There is one further ramification of variational principles in physics that should be mentioned here. As we have seen, the Principle of Least Action serves to identify the Newtonian equations of motion of particulate systems in fields of impressed force with the necessary condition (the Euler-Lagrange equations) for stationarity of paths in configuration space under a certain cost functional (action). The trajectories (in phase space) of these equations of motion are such that the Hamiltonian (total energy) is constant on them; indeed, the entire circle of ideas we have described is irrevocably tied to conservative systems. In any case, a most significant insight into the relation between conserved quantities, like energy, and associated variational principles was provided by Emmy Noether in a famous theorem which appeared in 1918. This theorem in fact serves to relate three fundamental physical notions: invariance (conservation), optimality (stationarity), and *symmetry*. In simplest terms, Noether's Theorem can be stated as follows: Given a cost functional like

$$\rho(y) = \int F(x, y, y')dx$$

and a one-parameter group of transformations (symmetries)

$$\begin{cases} x = \Phi(x, y, y', \tau) \\ y = \Psi(x, y, y', \tau) \end{cases}$$

depending differentially on the parameter τ, which leave the cost functional *invariant*. Then there is an associated *conserved quantity*, maintained constant on the trajectories specified by the corresponding Euler-Lagrange equations. The converse is also true: if there is such a conserved quantity, there is a one-parameter group of symmetries which leave the cost functional invariant. We can actually write down the conserved quantity, explicitly in terms of the cost functional and the associated group (Gelfand and Fomin 1962).

These facts indicate a deep relationship between optimality considerations and such concepts as similarity and scaling. We shall see further connections in subsequent sections.

Finally, we have already alluded in the preceding section to the difficulty of extracting first-order differential equations from optimality conditions. Why this is so can be seen clearly, in the present context, from the Euler-Lagrange form 14.15 of the equations of motion. This is in general a second-order equation in the configurational variable x, which when fully expanded, is of the form

$$\left(\frac{\partial^2 L}{\partial v^2}\right)\frac{d^2x}{dt^2} + \left(\frac{\partial^2 L}{\partial x \, \partial v}\right)\frac{dx}{dt} + \frac{\partial^2 L}{\partial v \, \partial t} - \frac{\partial L}{\partial x} = 0.$$

It is possible to require that the Lagrangian L has the property

$$\frac{\partial^2 L}{\partial v^2} = 0$$

which does yield a first-order equation, without assuming that the Lagrangian L is actually independent of v (but rather, linear in v). This situation does indeed have interesting physical ramifications (see, e.g., Sundermeyer 1982), but clearly leads only to first-order systems of a very special type.

It is, on the other hand, possible to imagine *embedding* a general first-order system (dynamical system) into a *larger* system, whose trajectories are generated by optimality conditions. This leads naturally to the concept of a *control system,* which we shall now discuss briefly.

Optimality Principles in Engineering: Optimal Control

Engineering or applied sciences, while based on theoretical sciences, differ substantially in emphasis and outlook. For instance, in mechanics, the main problem is to characterize the configuration of a system, or some function of it, as a function of time, with impressed force held fixed. In engineering, the main problem is to express a system behavior (output) as a function not so much of time, but of the impressed force (input). In some ways, this is a more general approach.

In a continuous-time context (to which we shall restrict ourselves here, for simplicity) the starting point for the study of control theory is simply a set of rate equations of the form

$$dx_i/dt = f_i(x_1,\ldots,x_n,\ a_1,\ldots,a_m) \qquad (14.17)$$

where the x_1 are *variables of state,* and the a_1 are *parameters.* These look very much like the equations of motion of a mechanical system (cf.

equation 14.8 above), in which the variables of state are the analogs of the mechanical *phases,* and the f_i are the analogs of *impressed forces.* However, from a conceptual point of view, rate equations like 14.17 are much more general. For one thing, there is no separation of state variables into configurational and velocity variables; no associated notions of energy, work, or other mechanical variables which play such an important role in pure mechanics. Indeed, such systems violate from the outset the basic Newtonian presupposition that configurations and velocities can be freely chosen; in systems like 14.17 the velocities of the state variables are *determined* by the states. This makes all the difference.

Once the functions f_i are given, we can see that each vector $\alpha = (a_1, \ldots, a_m)$ of parameter values determines a corresponding tangent vector at every state. Thus, the space of parameter vectors can be identified with the tangent spaces to the states—i.e., with the *velocity* vectors of standard Newtonian mechanics. If we were to think of the state variables as analogous to *configurational* variables of a mechanical system, such a system would have m degrees of freedom if $m < n$. Choosing different parameter vectors α gives us the capacity to modify the system dynamics—i.e., to *steer* the system from state to state. For this reason, such modifiable parameters are usually called *controls.* The main problem of control theory is to move the system 14.17 from some initial state to a desired final state by appropriately choosing the controls as functions of time. Any controls $\alpha(t)$ which do this job can be called *admissible.* The problem of *optimal control* is to choose some admissible control $\alpha^*(t)$ which also minimizes or extremizes some cost functional, such as the total time (or cost) associated with the corresponding path from initial to final state.

The body of literature generally called *optimal control theory* is essentially a paraphrase of the Hamiltonian formalism for conservative mechanical systems. In a nutshell, this requires introducing auxiliary (generalized) "velocity" vectors $v_1, \ldots, v_n$, which depend on the controls $\alpha(t)$ and on the states (configurations), in such a way that the cost functional becomes the Lagrangian for the larger system. We can then solve for the optimal controls $\alpha^*(t)$ which make this work; there are standard algorithms for doing this, such as the Pontryagin Maximum Principle, or Bellman's Dynamical Programming. As noted, we cannot go into details here, but they can be found in a host of standard texts, as well as in the primary references alluded to earlier.

Several conceptual points should be stressed here. First, the aim of optimal control theory is to characterize a time series of control vectors—i.e., a *policy;* in the continuous-time case we have been discussing here, this is just a curve $\alpha = \alpha(t)$. Second, it must be emphasized that the *original* first-order dynamics 14.17 do not arise in general from any kind of optimal principle; only when the *optimal* control $\alpha^*(t)$ is imposed does the augmented system follow optimal paths. Third, the "controls" themselves are imposed by continuously *resetting initial conditions* for the "velocities," v_i, and *not by changing the force*. From the standpoint of mechanics, this force is manifested by the Lagrangian function, which is the integrand of the cost functional; this stays fixed throughout. Thus the term "forcing," which is often used to describe the elements $\alpha(t)$ of the control space, can be quite misleading.

Optimality in Biology

Biology has always seemed to sit apart from even the most powerful developments in the physical sciences. For one thing, since at least the time of Aristotle, biology has been strongly associated with *telos*—purpose, goal, end. Kant (who understood Newtonian mechanism very well) claimed that the essential features of the organic world were so bound up with *telos* that a mechanical explanation of them was impossible in principle. Indeed, the entire thrust of Newtonian mechanics was to abolish *telos* entirely from science; this is why Kant picturesquely stated that there could never be the "Newton of the leaf," who could do for a blade of grass what Newton had done for the solar system.

Furthermore, biology has no analog of the solar system itself; it has no pendulums or inclined planes or lodestones, which manifest fundamental phenomena in simple manageable situations. From the physical point of view, biology deals with material systems of which even the simplest is already unimaginably complicated. The physicist there gladly abandons biology to the biologist, trusting in the universality of his laws and the belief that only the sheer complication of organisms stands in the way of complete understanding of organic behavior on familiar physical grounds. For his part, the modern biologist has finally seen a way to make mechanism and biology overlap; it is called "molecular biology," and its more philosophically inclined practitioners have praised their own accomplishments in this area with words which remind us of Laplace and Lagrange at their most extravagant.

Let us return to the notion of *telos*. One explicit way of excluding *telos* from physics is by postulating that present change of state must never depend on future states (or future forces). To us, this still seems reasonable enough; it is easy to accept that such a dependence on the future is incompatible with traditional views of determinism. On the other and, it was early recognized that the variational principles of physics, which we reviewed above, seem already to violate this maxim; we need both a present and a future configuration to determine a path. Thus, the Principle of Least Action, say, which is at the very heart of theoretical mechanics, looks more telic than mechanics itself allows. This has always bothered people, and many have taken the trouble to rationalize it away on various grounds, which we need not pause to review here. But these facts point to a perhaps deep relationship between the nature of optimality principles in general and the things we do not understand about organic phenomena.

However, there is a more immediate relation between optimality and biology. In 1859, Darwin published his *Origin of Species*. The main thrust of this book was to build a case for evolution, through the postulation of an explicit mechanism whereby it could occur, namely, natural selection. Natural selection was quickly paraphrased as survival of the *fittest*, and thus evolution could be regarded as a universal process for generating optimal organic structures. "All" we need to do to make this precise is to specify "fitness" explicitly as a cost function(al) on an appropriate class of organic designs; we could then not only generate the fittest organisms on purely theoretical grounds, but also the entire evolutionary trajectories which produce them.

Indeed, the situation looks very much like the one we have already described in connection with Newton's Second Law. Instead of a Newtonian particle, with its configuration and velocity, we have an organism, with a collection of genetic and phenotypic qualities, each of which is evolving at some rate. Instead of an external world which imposes forces, we have externally imposed selection mechanisms, whose effects are determined by the qualities on which they act and the rates at which these qualities are changing. One would imagine that after a century and a quarter, there would be an elaborate body of theory built up on this basis, especially in light of the importance of the associated ideas within physics itself, and the strong impetus toward relating biology and physics. However, nothing of the kind is true; except for a few special limiting cases, which we shall discuss below, and despite intense but amorphous speculation by biologists over the years on the nature of selection mechanisms, there is no such theory.

Let us briefly sketch one application of such a theory. We have already mentioned Noether's Theorem, which deals with groups of transformations (symmetries) leaving the cost functional invariant. In such situations, the theorem tells us that there is always an associated invariant—a quantity which is constant on the system trajectories in phase space. This means that the trajectories themselves are restricted to lie on hypersurfaces of codimension 1 in phase space. Each value of the invariant determines a different such hypersurface. Now the *paths* in *configuration* space, on which the cost functional is defined, are projections of these phase-space trajectories onto configuration space. This means that each path is the image of many trajectories, one from each of these hypersurfaces. However, the value of the cost functional (action) along a path will be different, depending on which hypersurface we are projecting from; roughly speaking, *the time it takes to traverse the path* depends on the value assigned to the conserved quantity, and hence on the particular choice of initial conditions.

From this, it follows that any transformation of phase space which leaves the *paths* invariant will necessarily be *optimality preserving*, in the sense that the transform of an optimal trajectory is again optimal. Such a transformation must map invariant hypersurfaces onto invariant hypersurfaces, and trajectories onto trajectories; basically, it serves as a *scaling transformation* between different energy levels (i.e., it serves to compare behaviors on distinct hypersurfaces). Such transformations are more general than those considered in Noether's Theorem, in which the hypersurfaces themselves are left invariant.

A *group* of such transformations will have as its orbits a field of manifolds cutting each of the hypersurfaces obliquely. If the group is large enough, these orbits will be one-dimensional; any point on one hypersurface will have a unique "corresponding point" on any other hypersurface. A vestigial example of this situation is well known in thermodynamics under the rubric of the Law of Corresponding States (see Rosen 1978a).

Such a situation can have many evolutionary implications. In that context, the application of one of our transformations to a point in phase space can be looked upon as a "mutation"; it is essentially a particular way of changing initial conditions. The "mutated" situation will then determine a new trajectory, which is obviously again optimal, and which can be computed from the original "unmutated" trajectory by means of the very same scaling transformation. One immediate corollary of this circle of ideas is the familiar Theory of Transformations of D'Arcy Thompson, which says that "closely related species differ phenotypically only by a rescaling (i.e., are similar)." However, to interpret such results explicitly, in particular evolutionary contexts, requires close attention to what is meant by genotype (species) and by phenotype in each particular case (see Rosen 1984a for detailed discussions of these matters).

Obviously, a thoroughgoing theoretical treatment of the relation between optimality and evolution is long overdue. Moreover, because of the parallels which have often been drawn between evolution and morphogenesis, between evolution and learning, and between biological evolution and corresponding processes in social systems, any theoretical developments in the understanding of biological evolution will radiate in many directions. If the history of optimality in physics provides any guide to the future, it is to expect precisely this.

At present, however, the best-developed applications of optimality to problems of biological structure and function occur at a different level. In essence, they pertain to situations in which, intuitively, selection pressure is so strong that we may suppose organisms already to be optimized. Moreover, the cost functions with respect to which organisms are optimal are patent from engineering principles. The first, to my knowledge, to explicitly enunciate such a principle was N. Rashevsky (1960), who extrapolated it from a long series of studies of organic form and function. He stated it as follows: "For a set of prescribed biological functions of prescribed intensities, an organism has the optimal possible design with respect to economy of material used and energy expenditures needed for the performance of the prescribed functions."

One immediate physiological system which satisfies the above conditions is the mammalian cardiovascular system. Selection is extremely strong on this system; even a small deviation from nominal design will clearly be eliminated very quickly. Further, the cardiovascular system is basically hydraulic in character, and considerations of fluid flow are bound to be important. On the other hand, the system is one of incredible complication, consisting typically of many thousands of miles of conduits, permeating to within a few microns of each of the 10^{15} cells which comprise a mammalian body. Nevertheless, many of the most intimate details of vascular architecture and organization follow relatively quickly from straightforward optimization with respect to resistance in the fluid path, economy of materials, and minimization of turbulence. (For details and a review, see, e.g., Rosen 1962b.)

On an entirely different level of biological organization, M. Savageau (1976) has considered the organization of biosynthetic pathways in cells, especially bacterial cells. Certain features of these pathways seem quite invariant, which is surprising considering the plethora of apparently equivalent alternatives that are available. For instance, even in straight-chain pathways, we invariably find that (a) the end-product of the pathway acts as an inhibitor of the initial step; (b) no intermediate step inhibits the initial step; (c) the end-product does not inhibit the intermediate steps. Why always this pattern, out of all of the alternative $2^{n(n+1)/2}$ possible patterns in an n-step pathway? Savageau shows that the actual pattern in fact simultaneously optimizes pathway performance with respect to a number of apparently distinct cost functions, such as the overall sensitivity of the pathway, overall gain, and sensitivity to exogenous end-product. And here again, small variations in pattern will cause large changes in overall fitness, so that we may expect selection to act strongly.

Numerous other examples could be adduced to show that optimality considerations already play a substantial role in the theory of biological activities. But I am convinced that they will be absolutely indispensable in years to come.

Optimality in Biomedicine

All technologies proceed from an underpinning of theoretical science. Numerous technologies arise from the various facets of biology: old ones like medicine, agriculture, and husbandry, to name a few, and new ones like genetic engineering and bioengineering. This list is bound to grow, and as it does, it will transform our society in ways we cannot yet imagine, as the "industrial revolution" did previously.

All the biologically based technologies involve ideas of optimality in two distinct ways. First, just as in physically and chemically based engineering technologies, the main point is to exert control upon some system of interest in the most effective way. But on another level, the systems we wish to control are now themselves generally control systems. This last fact introduces a new element into the biologically based technologies, as we shall see.

Almost any biotechnology would serve to illustrate the points we wish to make. For instance, the generation of optimal management and harvesting strategies for various kinds of crops is an endless and fascinating field and of the greatest importance for all of us. But for brevity, we will restrict ourselves to the bundle of technologies we collectively call medicine.

The field of medicine can be subdivided in various ways. On one hand, we find *hygiene,* which has to do with maintenance or preservation of a "state of health." When a deviation from this "state of health" occurs, we must on the one hand find out what the deviation is (*diagnostics*), and then generate, insofar as possible, a strategy for the restoration of health (*therapeutics*). Where this cannot be done, auxiliary supporting systems to simulate health (*prosthesis*) must be fabricated.

In what follows, we shall restrict ourselves to therapeusis, which is typical, and which overlaps the other divisions sufficiently to illustrate their problems as well.

As noted above, the role of therapeutics is to move a patient back to a "state of health," from his present state, by means of a suitable *therapy.* We assume that diagnosis is capable of telling us at least something about his present state; thus a therapy is a control function which will move a patient from his present state to a desired state. We also want to choose this therapy, or control, in such a way that it is in some sense optimal. In other words, therapeusis is an exercise in optimal control theory, as Bellman early recognized (Bellman, Jacquez, and Kalaba 1960).

The problems of therapeutics are thus like those familiar from physically or chemically based control problems—e.g., steering a rocket. However, as we noted earlier, biology confronts us with classes of problems which are refractory to resolution on grounds of classical physics alone. Thus, before we can begin to approach therapeutic problems in biomedicine as problems in optimal control, we have some serious technical and conceptual problems to face.

We stated at the outset that to define an optimality problem, we must have at hand a universe of discourse U, consisting of all potential (admissible) solutions of a problem, and a cost function ρ associating a number (cost) with each admissible solution. We also stated that the "optimal solutions" are extremely sensitive to the specifications of both U and ρ. In standard physical applications of optimality, the characterization of both of these is relatively straightforward, as we have seen. But in biomedical applications, they are almost never so. They involve not only deep questions of theory, but also empirical problems relating to system identification and state identification.

In engineering applications, for instance, it is not hard to decide what shall constitute a *state* of the system we wish to control. It is not hard to specify the dynamical laws governing change of state; in these situations, there is standard general theory which may be brought to bear. It is usually not hard to characterize the desired state we wish to reach nor to find what state the system is actually in at an instant. Thus, the universe of discourse U, the set of admissible controls, is in principle well determined; that is, we can in principle answer the question "is this given function $\alpha(t)$ an admissible control or not?" with a yes or no.

But in the biomedical area, each of these steps presents us with unsolved problems. How shall we characterize the "state" of an organism which we desire to control? How are the "states" changing under the influence of time, and of ambient forcings? How can we find out what "state" an organism actually occupies at an instant? And how shall we characterize the "healthy states," those which must be the target of therapy? Until we answer these questions satisfactorily, our universe of discourse is uncharacterized, and in formal terms we cannot even formulate a cost function, let alone find its extrema.

Sometimes, of course, we can answer some of the above questions. We can, for example, associate a "state of health" with a situation in which the values of certain system variables (i.e., a body temperature, or a blood pressure value, or a glucose or cholesterol concentration) fall within preassigned norms. When measured values depart from the corresponding norm, then by definition the organism is no longer in a "state of health," and the target of therapy is to bring the measured values back to their nominal values. In other situations, a "state of health" may pertain to more qualitative characteristics: the shape of an electrocardiogram or electroencephalogram tracing, the degree of enlargement of an organ, or the presence of adventitious masses. Here, the ascertainment of the "state of health" is more a matter of pattern recognition than of numerical measurement.

But even if all these matters could be disposed of, we must still write down the dynamical laws governing the system to be controlled—the manner in which the system states change in time. This is, after all, the basic ingredient in deciding whether a control is admissible, and thus in defining the universe of discourse U. The only way to do this is to have recourse to a *model* of the system; in distinction to the engineering situation, such a model cannot be based directly on general laws, but must be to some degree conjectural. Once a model is adopted, the problem of admissibility becomes resolvable *relative to that model*, but *only* relative to that model.

This brings us back again to basic science, and to the other basic role that optimality considerations play in the biomedical enterprise. For we naturally want to have the "best" model as the starting point for the generation of optimal therapies. The fact is that in the biotechnologies, to a far greater extent than in other technologies, the modeling enterprise, and the theory which underlies it, is directly involved in the "practical" problems of generating optimal strategies from the very outset.

The choice of cost function in biomedical applications also involves modeling in an essential way. Basically, this is because the old idea of the "magic bullet," the therapy without side effects, or benefit without cost, has had to be abandoned. Almost by definition, a side effect is something which cannot be predicted on the basis of a direct control model of the type we have been considering; instead, we must pass to a larger model in which a particular side effect becomes a direct effect of imposing control (see Rosen 1974). A cost must be assigned to each of these side effects, and incorporated into the appropriate cost function, which is chosen so as to measure benefit of therapy against cost of side effects. Alternatively, instead of modeling these side effects explicitly, we can model them indirectly, by imposing some kind of probability distribution on our original problem; this leads to questions of generating strategies in the presence of uncertainties, at which Bellman was so adept.

Space precludes considering these matters in any greater detail. At this point, the reader must be referred to the literature for further discussions and examples, as for example the review volume of George Swan (1984) and the multitude of references contained therein.

We cannot do better in concluding this necessarily brief and sketchy review of the role of optimality in biology and biotechnology than quote some words of Bellman (1968), which sum up his view of the fertility of this area, and of the role of mathematics within biology:

> If you do believe that a vital field of science is a rich source of mathematical problems, then the field of the biosciences is for you. Furthermore, it is the last frontier. The field of physics will never be completely worked out, but the returns these days seem marginal compared to the money and manpower expended. However, the field of biology is so rich and complex that one cannot visualize its being exhausted any time in the next hundred or two hundred years.

> Thus the conclusion we reach is that . . . research in the biomedical domain is the activity for the young mathematician. He will be delving in areas replete with hundreds of new and fascinating problems of both interest and importance, where almost no mathematicians have ever set foot before.

> It is sad to see brilliant young people scrambling after crumbs when banquet tables are waiting in the mathematical biosciences.

> (pp. 127–128)

We hope to have indicated where some of these banquet tables are located in the above discussion, and some idea of what is on the menu. And everything is à la carte.

❧

CHAPTER 15

Morphogenesis in Networks

Basic Ideas

One of the most ancient, and at the same time most current, fields of theoretical biology is that concerned with morphogenesis—the generation of pattern and form in biological systems. This chapter is devoted to the development of an integrated framework for treating morphogenetic problems, not only because they are of the greatest interest and importance in their own right but also because they tell us some important things about theoretical biology in general, and they help us articulate the position of biology vis-à-vis other scientific disciplines. I will first discuss theoretical biology in general, and then morphogenesis in particular.

The basic question with which theoretical biology is concerned is simply, What is life? (see chapter 1). Our intuitions tell us that there are the sharpest distinctions separating the organic from the inorganic, and these intuitions provide the basis for the study of biological phenomena as an independent discipline. Yet these intuitions have thus far successfully resisted every attempt at formal characterization. As a result, the subject matter of biology as a whole cannot be explicitly defined.

Ever since the time of Newton, physics has claimed jurisdiction over every question that can be asked about the behaviors of material systems. Newton and his successors showed how to understand the behaviors of any particulate system. It is difficult for us to appreciate today just how profound a revolution this development has wrought on the way we think about the world. By explaining planetary motion on the basis of a few simple laws, Newtonian mechanics destroyed all vestiges of the animisms that characterized earlier times. Moreover, in combination with the corpuscular theory of matter, which holds that every material system is composed of particles, mechanics provided a universal strategy for approaching all problems about such systems, including, of course, organisms. Namely, all we need to do is to characterize the constituent particles of such a system, identify the forces acting on them, and all else will follow. Such ideas accord well with the old Cartesian view of the organism as an automaton (i.e., a mechanism). So we see here, in a nutshell, the modern program of molecular biology. According to this view, solving organic problems is merely a technical matter, raising no new considerations of principle; biology has been reduced to physics, in thought if not yet in fact.

An immediate corollary of this viewpoint is that there is no difference at all between organisms and any other kind of physical system. As far as physics is concerned, any perceived differences are differences in degree and not in kind. Assertions to the contrary are dismissed as vitalistic, especially since it has been repeatedly shown that the matter of which organisms are composed is just like all other matter, and the forces that move this matter are just like any other forces.

Nevertheless, it is not only possible but necessary to maintain that there is something special about organisms, without at the same time embracing a crude vitalism. What is clear is that, whatever it is about organisms that makes them special, it cannot be characterized in terms of physical laws alone. For there is nothing at the level of material forces that distinguishes between living systems and nonliving ones. Indeed, nothing in the laws of physics implies that organisms are necessary, or even that they exist. Where, then, shall we look for this hypothetical characterization?

What makes organisms special are the homologies between their behaviors. Organic systems are in some deep sense *similar* to each other, and this similarity is maintained despite the fact that different organisms may from a physical point of view have hardly a molecule in common (i.e., they may be composed of entirely different particles) and the forces between these molecules are correspondingly different. From physics we learn that behavior is determined completely by constituent particles plus forces; in organisms we find the widest variations in these constituents, and yet we find profound homologies between the things that organisms do. This is but one of the several nasty corners into which an uncritical acceptance of reductionistic dogmas in biology drives us.

The behavioral homologies between organisms form a basic part of our intuitions about living systems, which suggests an alternative strategy for characterizing and studying organisms. Instead of starting from the notion that an organism is a physical system (as the physicist would) and attempting to work up from its constituent particles to its gross behavior, let us instead try to characterize the behavioral homologies first, and then work back down to physical structure. In this procedure, we treat relations between organic behaviors as primary, in the same way that a physicist would treat physical structure as primary—i.e., we treat these relations as *things*. This approach, which turns the reductionist program upside down, was initially developed by Rashevsky, who coined the term *relational biology* to describe it. Its essential feature is that it begins from the hypothesis that organisms are basically alike and concentrates on characterizing what it is that makes them alike. (See chapter 17 for a fuller description.) Physical approaches, on the other hand, begin with the premise that organisms are not alike and concentrate on the differences between them.

The relational approach is congenial to mathematicians. To an overwhelming degree, mathematics is concerned with the structures relating different mathematical objects. For instance, an algebraic structure is characterized as a group, not in terms of the nature of its elements (which may be anything), but rather because of certain global features that it shares with all other groups. These global features are captured abstractly in terms of the familiar group axioms, which impose a common abstract structure on all groups, regardless of the nature of their elements. Different groups may be compared by means of structure-preserving mappings (morphisms), which maintain that structure. Accordingly, the study of group structure per se can be carried out in a higher-level mathematical structure called a category. This is the way relational biology also proceeds: beginning from an abstract characterization of organization (which gives a category), and proceeding from the abstract to the concrete by a process of realization or representation.

Morphogenetic processes provide a rich subclass of biological behaviors. However, pattern generation is not restricted to biological systems; a pattern-generating process may be realized in many other contexts as well. By exploiting this, we shall find a different, nonreductionistic way of relating biology to other sciences (for instance, to physics) and see how we can construct physical models for biological processes—the model being related to the process because they both represent alternative realizations of the same abstract relational structure.

Biologists have been concerned with problems of organic form and its generation for a long time, because the form of an organism is one of its most immediate and obvious characteristics. Form provided the foundation for establishing the taxonomy of Linnaeus, which marked perhaps the beginning of biology as a strict science and led almost inevitably to the idea of biological evolution. (Taxonomy, the study of classification, is regarded by most scientists as a rather pedestrian activity. However, it is the basic step in establishing any science; think of the role of the periodic table in chemistry, and of the fact that most of the deep problems within mathematics itself are classification problems).

Reductionistic approaches have had little success with problems of organic form and its generation. Indeed, the current preoccupations with molecular biology have made these problems harder rather than easier. For instance, the *gene* was originally regarded, by Mendel and the early geneticists, as a hereditary unit that governed form in an organism. According to this early view, the genotype, or totality of genes in an organism, controlled the organism's phenotype, or form. However, beginning with the Beadle-Tatum experiments in the early 1940s, and culminating in the Watson-Crick model for DNA and the subsequent elucidation of the mechanism for protein synthesis, the role of the gene has been completely reinterpreted. It is now regarded as a molecule (DNA) that modulates *intracellular* chemistry, through controlling the synthesis of specific catalysts or enzymes. This picture is infinitely far removed from the original view of a gene as a unit governing gross morphological features in a whole multicellular organism. We thus find an enormous chasm separating what the gene was originally supposed to do and what we now believe it actually does; the chasm is the distinction between chemistry and geometry, and, from a purely reductionistic viewpoint, it is simply unbridgeable.

Let us turn now to the definition of a class of abstract relational structures that will allow us to study many (perhaps all) morphogenetic or pattern-generating processes in a common formal framework. Through appropriate modes of realization, we may apply these structures to an enormous gamut of morphogenetic problems, spanning many organizational levels. To list a few: the spontaneous folding of a newly synthesized polypeptide chain into an active, three-dimensional, folded structure; the assembly of virus particles and other cellular organelles from their molecular constituents; the phase transitions of ferromagnets and other thermodynamic structures; the firing patterns of neural networks; the breathtaking phenomena of embryology and development; the distribution of species on a landscape; and many others besides.

We view morphogenesis as a dynamical process occurring in populations of spatially distributed units. Such a population may be described by the following data:

1. A set $N = \{n_i\}$, $i \in I$.

The set N is our population; the individual elements in the population are the n_i. I is simply an index set, but it should be thought of as a set of spatial coordinates, so that the element n_i is intuitively regarded as the element or unit of our population sitting at position i. We make no initial restriction on N: it may be finite or infinite. Likewise, I may be finite, infinite, discrete, or a continuum.

2. A set S.

We shall think of S as a set of "internal states" for the elements n_i in our population. It is no real restriction to suppose that all the elements in our population have the same set of internal states. If we assign a copy of this state set S to each n_i in N, we obtain a picture very much like a fiber space, where N is the base space and S is the fiber.

We can now define a *pattern:* it is obtained by assigning to each element n_i in our population a particular state in S. Mathematically, a pattern is simply a mapping $f : N \rightarrow S$. The set $H(N, S)$ of all such mappings is the set of all patterns. Any particular pattern can thus be regarded as a cross section of the fiber space we have defined.

To talk about pattern generation, we must introduce some dynamical ideas in the space of patterns $H(N, S)$. First, we need a time frame. Thus we need the following:

3. A linearly ordered set T.

This set T will be the set of *instants*. In practice, we will always choose either $T = \mathbb{Z}$ (integers) or $T = \mathbb{R}$ (real numbers). The former specifies *discrete time;* the latter is *continuous time.*

Now we want to discuss pattern generation in a local way. That is, we want to know how the state of each element n_i in our population is changing in time. We want this local change of state to depend on (1) the state of the element itself at an instant; (2) the states of other elements in the population at that instant; and (3) possibly the values of external environmental parameters at that instant. To do this, we must first characterize those other elements in N whose states affect the local change of state of n_i; we shall call these elements the *neighbors* of n_i. This can be done succinctly by defining the following:

4. A mapping $U: N \rightarrow P(N)$.

Here, $P(N)$ is the power set of N, the set of all subsets of N. The image $U(n_i)$ of each element n_i in N is the set of all the neighbors of n_i. We require only that n_i be an element of $U(n_i)$ for every element n_i. Intuitively, if n_j is a neighbor of n_i, then its state at an instant helps determine the change of state of n_i. In biological terms, we would say that such an n_j is afferent to n_i.

Finally, we need to specify the rule, or dynamical law, that governs how each element n_i changes state from instant to instant. This will be called a *local state transition rule,* or LSTR. We recognize two cases, according to whether time is discrete or continuous:

5a. Discrete-time LSTR:

$$n_i(t + 1) = \Phi[U(n_i)(t), \; \alpha(t)].$$

Here we have written $n_i(t) =$ state of the element n_i at the instant t; $U(n_i)(t) =$ states of the neighbors of n_i at time t (where we recall that n_i is always a neighbor of itself), and (t) is a vector of values of external parameters, if any, that affect the local state transitions.

If time is continuous, and if N is a manifold, we modify the preceding accordingly and define the following:

5b. Continuous-time LSTR:

$$dn_i(t)/dt = \Phi[U(n_i)(t), \; \alpha(t)].$$

These data define a class of abstract mathematical structures that we shall call *morphogenetic networks*. Clearly, given any initial pattern f in $H(N, S)$, the LSTR defines a pattern-generating process in $H(N, S)$, defining a new pattern at each instant by telling us how each individual element n_i changes state from instant to instant.

Notice that we are not allowing the external parameters α to depend on spatial position, but only on time. If we allowed α to depend on space as well as time (i.e., if we took $\alpha = \alpha(i, t)$), we would thereby convert the environment into a *template*. Intuitively, this would mean that the morphogenetic processes we consider are not determined by autonomous interactions between the elements n_i, but require information supplied from the environment through the spatial dependence of parameter values $\alpha(i, t)$. Such template-assisted pattern generation is important for many purposes, and we can easily incorporate it into our considerations, but we shall not consider it directly. By excluding such template-assisted processes, we shall restrict ourselves to situations of *self-organization*, in which the generation of patterns is governed essentially by the character of the interacting elements alone.

Examples

Ising Models

In physics, the closest analog to biological pattern generation or morphogenesis is perhaps the *phase transition*, in which an abrupt change of properties (e.g., from gas to liquid) can occur under appropriate conditions. A favorite approach to the study of such behavior lies in a class of model systems called *Ising models*, which can be represented as morphogenetic networks. This development shall for simplicity be restricted to two dimensions, but it can be generalized to any number of spatial dimensions.

First, we need to specify a set N of elements, an index set I of spatial coordinates, a state set S, a set of instants T, and a local state-transition rule. Let us proceed as follows:

1. $I = \{(\mu, \nu)\}$, where $\mu, \nu = 0, \pm1, \pm2, \ldots$
2. $N = \{n_i\}$, $i \in I$

3. $S = \{+1, -1\}$

4. $T = \mathbb{Z}$

5. $U(n_{\mu,\nu}) = \{n_{\mu,\nu+1}, n_{\mu,\nu-1}, n_{\mu+1,\nu}, n_{\mu-1,\nu}\}$

6. $n_{\mu,\nu}(t + 1) = 1$ if and only if

$$\sum_{n_k \in U(n_i)} [1 + n_k(t)]^2 + \alpha \leq \sum_{n_k \in U(n_i)} [-1 + n_k(t)]^2 - \alpha$$

otherwise $n_{\mu,\nu}(t + 1) = 0$.

In words, condition 1 says that the index set I consists of points in the plane with integer coordinates. Condition 3 says that our elements can take on only two internal states—i.e., they are like switches. Condition 4 says that time is discrete. Condition 5 says that only nearest-neighbor interactions are allowed. Condition 5 says that the state of any element n_i at time $t + 1$ is the state that minimizes the expression

$$E_i(t) = \sum_{n_k \in U(n_i)} [n_i(t + 1) - n_k(t)]^2 - \alpha n_i(t)$$

where α is a parameter; $E_i(t)$ can be thought of as a "local free energy" and α as an external field. It is easy to see that at low field strength, rule 6 says that each element will go to the state opposite to the majority of its neighbors at the preceding instant; if α is large enough, each element will follow the field, independently of its neighbors.

Thus when α is small, the network will tend asymptotically toward a checkerboard pattern of alternating positive and negative states. When α is large, all states will be negative. This in effect is the phase transition.

These simple networks can be generalized in various ways. We can generalize them to three or more spatial dimensions by appropriately changing the index set I. We can take a more general state set S; we can allow it to contain $r > 2$ elements, or even let it be a continuum. We can allow more complicated interactions than those arising from nearest neighbors, and of course we can modify the local state-transition rule. Seen in this light, the simple Ising models of the physicist are in fact very special cases of the kinds of excitable networks found in biology— for example, genetic control networks and neural networks.

One other feature of the simple Ising models, worth noting here, is the *interpretation* we give to a local state transition $+1 \to -1$ or $-1 \to +1$. The most obvious interpretation is to regard such a state transition as a kind of *differentiation* of a single element fixed in space. However, there are two other interpretations that are of interest:

1. Think of $+1$ and -1 as the names of two different kinds of elements, rather than as states of a single element. In this case, a transition $+1 \to -1$ can be regarded as the movement of the original element $+1$ away from its original position, and the migration of an element of type -1 into that position. That is, we can regard our elements as mobile, and the local state transition rule as governing their movement in space. The Ising models then become models of *morphogenetic movement;* in physics they thus become models of such things as phase separations rather than phase transitions.

2. Think of $+1$ as denoting the *presence* of an element at a position in the network, and -1 as the absence of an element. Then, the local transition $+1 \to -1$ means that an element has "died" at the position; the reverse transition $-1 \to +1$ means that an element has been "born" there. Under this interpretation, the Ising networks become models for *differential birth-and-death processes.* Such networks are studied in physics under the rubric of lattice gases.

These three interpretations represent in biology the three basic morphogenetic mechanisms: differentiation, morphogenetic movement, and differential birth-and-death. We can study such mechanisms within a common framework, and with a common formalism, simply by modifying our interpretation of the formalism. This unification is one of the most important features of the network formalism we have introduced.

Reaction-Diffusion Networks

The second example is a class of networks that can collectively be called *reaction-diffusion networks.* These can be thought of as arrays of elements ("cells") that possess some internal dynamics (e.g., chemical reactions) and that can interact with their nearest neighbors through diffusion of reactants. In such networks, we usually suppose that time is continuous ($T = \mathbb{R}$), and that the state set S is a manifold, whose points are represented by *n*-tuples $(x_1, \ldots, x_n)$ in some system of local coordinates. Thus the state of a particular element n_i at an instant of time is represented by the *n*-tuple

$$n_i(t) = [x_{i1}(t), \ldots, x_{in}(t)].$$

Under such circumstances, a local state-transition rule for an element $n_i(t)$ specifies how each of the coordinates $x_{ik}(t)$ is changing in time. Let us write the following:

$$dx_{ik}/dt = \phi_k[x_{i1}(t), \ldots, x_{in}(t)] + D_k \sum_{n_j \in U(n_i)} (x_{ik} - y_{jk})$$

for each $k = 1, \ldots, n$. The first term in such an expression describes the autonomous dynamics of the element (i.e., the reaction part), while the second term represents the interaction between neighboring elements (the diffusion part); the form I have chosen for this term is ordinary Fickean diffusion.

In the preceding example, I chose the index set I to be a finite, discrete set. In certain applications, it is convenient to let I itself be a manifold. In this case, the local state-transition rules become systems of partial differential equations (whose specific formulation is left to the reader), and the Fickean diffusion term is replaced by Laplacian operators (where we differentiate with respect to local coordinates in I).

Despite the difference in technical detail, note how closely the reaction-diffusion networks resemble the Ising networks considered previously.

Excitable Networks

As a final example, we shall consider two classes of what may be called excitable networks—networks of threshold elements that may be thought of as neurons. The first class includes the two-factor networks, originally introduced by Rashevsky (1960).

We begin with a discussion of the elements of such a network. In macroscopic terms, such an element is to be regarded as a switch—i.e., at any instant of time, the element is either *off* or *on*. These two macrostates will in turn be regarded as equivalence classes of underlying *microstates.* We will regard these microstates as comprising the first quadrant of the Euclidean plane; that is, a microstate is represented as a pair (x, y) of non-negative real numbers. We shall further suppose that these microstates change in time according to the following rules:

$$dx/dt = b_1 I - \lambda_1 x$$
$$dy/dt = b_2 I - \lambda_2 y.$$

Here, I represents an incoming "excitation" (ultimately to be derived from other elements in a network); the b_i are arbitrary numbers, and the λ_i are positive.

Now we define the "on" macrostate as the set of all microstates (x, y) for which $x \geq \theta y$, where θ is a number characteristic of the element, called its *threshold*. The "off" macrostate consists of all other microstates.

Such excitable two-factor elements can be organized into networks. The interaction between two elements in such a network is via macrostates, mediated by the microscopic state-transition rules already described. Thus if n_i is an arbitrary element in such a network, then at any instant of time t it will be in microstate $(x_i(t), y_i(t))$ and in macrostate $\sigma_i = 1$ or 0 depending on whether $x_i(t) \geq \theta_i y_i(t)$ or not. We then write the following:

$$dx_i/dt = \sum_{n_j \in U(n_i)} b_{i1} \sigma_j - \lambda_{i1} x_i$$

and

$$dy_i/dt = \sum_{n_j \in U(n_i)} b_{i2} \sigma_j - \lambda_{i2} y_i$$

which comprise the local state-transition rules for the network. Note that the neighborhoods $U(n_i)$ are no longer necessarily defined in terms of spatial contiguity; instead, we shall say that $n_j \in U(n_i)$ means that n_j is *afferent* to n_i, or that there is an "axon" running from n_j to n_i. We also note that the equations describing the local state-transition rules, which look deceptively linear, are in fact highly nonlinear, owing to the nonlinear relation of microstates to macrostates.

The two-factor networks are continuous-time systems. They are closely related to a class of discrete-time networks, called variously *neural networks* or *finite automata*. Indeed, it requires only minor technical modifications to pass from the two-factor nets to the discrete neural nets. Primarily, we must replace continuous time $T = \mathbb{R}$ by discrete time $T = \mathbb{Z}$. Next, we must drop the apparatus of microstates and consider that each of our elements is simply a switch, which at any discrete instant is either "off" (i.e., in state 0) or "on" (i.e., in state 1). Neighborhoods are defined as before. The simplest local state-transition rule in such cases is obtained by writing the following:

$$n_i(t + 1) = 1 \quad \text{if} \quad \sum_{n_j \in U(n_i)} b_j n_j(t) \geq \theta_i,$$
$$= 0 \text{ otherwise.}$$

Such neural networks, and some of their immediate generalizations, have played an essential role in such diverse areas as the modeling of brain function, the control of gene activation, the theory of computing machines, and the foundations of mathematics itself. In fact, the simple Ising models previously described are very special kinds of excitable neural networks.

These simple examples show what an enormous gamut is spanned by the concept of a morphogenetic network. A multitude of specific pattern-generating processes, in biology, physics, and elsewhere, fall into this framework and are hence related to each other. For instance, the spread of excitation in the brain is not usually regarded as a morphogenetic process, but both excitation and morphogenesis can be described in common terms. The result is not only an enormous conceptual simplification but also a great economy of representation.

Here is an open problem suggested by the foregoing examples. The excitable networks described had a rather simple individual element (basically, a switch) that could be included in networks of arbitrary geometry. On the other hand, the reaction-diffusion nets had arbitrarily complicated individual elements, organized into networks with a rather simple (nearest-neighbor) geometry. The problem is, what are the relative roles of geometry and element complexity in pattern generation? Can we always compensate, in some sense, for a simple geometry by making the interacting elements more complicated? Conversely, can we compensate for simple elements by allowing a more complex geometry in the networks? In biological terms, are there things that brains can do, by virtue of their rich global geometry, that could not be done in any other way?

Applications to Biological Pattern-Generation Problems

Here are a few examples of how the morphogenetic networks have been applied to particular morphogenetic problems in biology.

Differentiation

The cells of a mature multicellular organism constitute a clone—a population of cells derived from a single cell (in this case, a fertilized egg or zygote) by a genetically conservative process of cell division. In a complex organism such as a mammal, we can find many hundreds of distinct cell types, which if considered in isolation would appear entirely unrelated; there is as little in common between, for example, a neuron, a leukocyte, and a fibroblast, in morphology, chemistry, and physiology, as there is between *Amoeba* and *Paramecium*. A typical mammal contains about 10^{15} cells, so all the cell types are produced during only about fifty rounds of divisions on the average ($2^{50} \cong 10^{15}$). Understanding differentiation, the way it is controlled and initiated, is perhaps the major conceptual problem in development.

Clearly, to attack biological differentiation in all of its complexity would be foolhardy. Thus biologists have long sought various kinds of *model systems*—biological and nonbiological systems that manifest one or another facet of behavior that can be related to differentiation, but in a simplified context that allows detailed study. Among biological models of differentiation are studies of bacterial adaptation and of forms such as the cellular slime molds. The problem is compounded because it is hard to know just how far we may extrapolate from such model systems to true differentiation.

A theorist, on the other hand, would focus on the *simplest* phenomenon that could be considered a true differentiation, and seek a mechanism whereby it could be modeled and understood. Such a phenomenon is presented by the autonomous generation of biochemical and morphological *gradients* and *polarities*. These are true differentiations and in fact are the earliest differentiation events that occur in multicellular development. If we could understand how polarities and gradients can be autonomously generated, we would take a long step toward understanding differentiation in general.

How shall we autonomously generate a gradient in a system that initially does not possess one (i.e., that is initially homogeneous)? This is especially vexing since in familiar inorganic systems we know that gradients tend to disappear, being autonomously destroyed by diffusion. That is, in inorganic systems, gradients are *unstable* and tend toward a homogeneous stable situation.

An answer to this dilemma was suggested in 1940 by Rashevsky. The answer is simple, almost trivial, in hindsight, but at the time it was a most radical innovation: See to it that it is the *homogeneous* state that is unstable. Although this flies in the face of "physical intuition," it is in fact the only way a gradient or polarity may be autonomously established.

Rashevsky gave an example of how a homogeneous state could be unstable; basically, it involved a combination of diffusion with chemical reactions, which can create and destroy the diffusing species. If this is done properly, the joint effects of reaction and diffusion together can indeed generate gradients autonomously.

Rashevsky's example was rather complicated. A much simpler version of his approach was independently developed by the English mathematician A. M. Turing in 1952. I shall give a further simplified version of Turing's approach, show how it does indeed generate autonomous gradients, and then indicate how it leads directly to the class of reaction-diffusion networks described.

Consider a system composed of two "cells," as shown in figure 15.1.

Assume for simplicity that each cell can be described by a single state variable x_1, which we can think of as the concentration of some chemi-

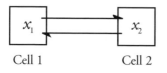

Cell 1   Cell 2

FIGURE 15.1

cal (Turing used the word *morphogen*); x_1 is the concentration of morphogen in cell 1, x_2 its concentration in cell 2. Assume further the following:

1. The morphogen is produced from an (infinite) external source and is catabolized at a rate proportional to its concentration.

2. The morphogen is free to diffuse between the two cells at a rate proportional to the concentration difference between them. Under these (rather weak) hypotheses, the dynamics of our two-cell system can be written as follows:

3.

$$
\begin{cases}
\dfrac{dx_1}{dt} = -ax_1 + S + D(x_1 - x_2) = (D - a)x_1 - Dx_2 + S \\
\dfrac{dx_2}{dt} = -ax_2 + S + D(x_2 - x_1) = -Dx_1 + (D - a)x_2 + S.
\end{cases}
$$

These are the deceptively simple (even linear) equations we need to analyze. It is easy to verify two things: First, that our system 1 has only a single steady state (x_1^*, x_2^*), given by

$$
x_1^* = x_2^* = S/a,
$$

which is *independent* of the diffusion or coupling between the cells. Moreover, this steady state is always homogeneous; the same amount of morphogen is present in each cell at the steady state.

Second, we can verify that the *stability* of the steady state is governed by the eigenvalues of the system matrix, which in this case is represented as follows:

$$
\begin{pmatrix}
(D - a) & -D \\
-D & (D - a)
\end{pmatrix}
$$

FIGURE 15.2

There are two eigenvalues:

$$\lambda_1 = -a, \ \lambda_2 = 2D - a.$$

The first eigenvalue, λ_1, is always negative. The second, however, is a function of D, which measures the strength of the diffusional coupling between the cells. If D is *small* (i.e., $D < a/2$), then the second eigenvalue will be negative as well. In this case, the steady state (x_1^*, x_2^*) is a *stable node,* and any gradients will autonomously disappear, as physical intuition teaches us to expect.

However, if D is large enough $(D > a/2)$, the eigenvalue goes positive. Now, instead of being a stable node, the steady state (x_1^*, x_2^*) becomes a *saddle point.* The trajectories in this case are indicated in the phase plot shown in figure 15.3.

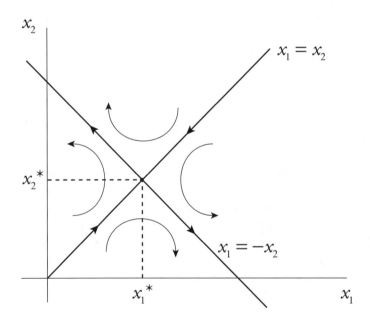

FIGURE 15.3

This system behaves physically as follows: If we initially start out in an *exactly* homogeneous situation $x_1 = x_2$, we will move to the steady state (x_1^*, x_2^*) as before. But if there is *any* initial difference between x_1 and x_2, that difference will grow autocatalytically. That is, we will be on one of the hyperbolic trajectories diverging further and further from (x_1^*, x_2^*) as time increases. Since x_1, x_2 are thought of as concentrations, they cannot go negative; the trajectory must stop when it hits an axis. But on the axes, one or the other cell is *empty*. All the morphogen has accumulated in the other cell. Thus in this case a true polarity has been autonomously and stably generated.

The system just described is essentially the simplest kind of reaction-diffusion network. It can be shown to satisfy the axioms of a morphogenetic network. It can be generalized indefinitely: more cells, more complicated neighborhoods, more morphogens, more complex internal reaction dynamics, more general diffusions. Not only are polarities and gradients generated here, of interest to embryologists, but also the *active transports* of materials against their diffusion gradients, which are ubiquitous in physiology.

The simple situation described is an example of a bifurcation phenomenon, analogous to classical problems such as the buckling of beams. In our case, the diffusion constant is the bifurcation parameter: as D passes through the critical value $(D = a/2)$, the initially stable steady state becomes unstable and bifurcates into two new stable steady states. Presently the bifurcation properties of more general reaction-diffusion nets are the subject of much study.

An entirely different way of attacking the problem of differentiation is from a genetic perspective. Differentiation poses as obstinate a problem for classical genetic ideas as it does for the physicist's belief in the stability of the homogeneous state: If the genome of a cell determines all of its properties, and if all cells of a multicellular organism contain the same genome, there seems to be no genetic basis on which the phenomenon of differentiation can occur.

A solution to this dilemma was posed by the French biologists François Jacob and Jacques Monod (1961) on the basis of their studies of bacterial adaptation. They found that genetically identical bacteria, grown in different kinds of media, synthesize only those enzymes required for metabolizing the nutrients in their own medium, and stop synthesizing all others. If switched to a different medium, such a bacterial cell will correspondingly switch its enzymic constitution. This phenomenon of bacterial adaptation resembles differentiation in that genetically identical cells can become biochemically different from one another (depending on their environmental histories); it is unlike differentiation in that it is reversible.

On the basis of their experiments, Jacob and Monod concluded that different adaptive states of bacteria were the result of *differential gene expression*. They postulated that the bacterial genome was replete with *control elements* ("operators") as well as genes specifying individual enzymes ("structural genes"). The states of these operators could be modified by environmental circumstances, and these operator states determined whether associated structural genes would be expressed or not (i.e., whether they were "on" or "off").

The combination of an operator with structural genes can be regarded as a formal unit (i.e., an *operon*), which is essentially a switch—with the same properties as a formal neuron. A network of such interacting units, which is the Jacob-Monod model for genetic differentiation, thus falls under the rubric of excitable networks—that is, of morphogenetic nets. The only novel feature here is that the neighborhood relation between operons is determined by chemical specificities of operator binding sites, rather than by tangible anatomic axonal connections; otherwise, the situations are formally identical.

The reaction-diffusion approach and the operon network approach represent two very different ways of attacking the differentiation problem. However, they can be integrated through the formalism of morphogenetic networks—for example, by placing the parameters of a reaction-diffusion network (rate constants, diffusion constants) under the control of genetic operators interacting through an underlying operator network.

Morphogenetic Movement

The prototypic experimental situation of generation of patterns through morphogenetic movement is called *cell sorting*. If tissue taken from a young embryo is disaggregated into its constituent cells and randomly reaggregated, the cells will begin to move relative to one another. In a relatively short time (24 to 48 hours), they will reconstitute at least an approximation of the original histology of the tissue. This is true even in organs such as kidney and lung, which have complex histologies involving many cell types.

The embryologist Malcolm Steinberg, who has extensively studied the role of morphogenetic movement in development, showed that cell sorting is a general phenomenon: it occurs even in artificial mixtures of cells that never come into contact in a normal embryo. He also suggested that the sorting phenomenon is what one would expect to find in mixtures of motile, differentially adhesive units. Thus he related cell sorting to physical phenomena of *phase separation,* such as the breaking of oil-water emulsions.

Thus a mathematical representation of sorting phenomena based on an Ising-model approach is immediately suggested. In such an approach a local change of state is to be interpreted as a morphogenetic movement—a migration of a unit away from its previous coordinate and its replacement with a new (generally different) unit. In these circumstances, however, we must impose an important constraint: If we assume no changes in the numbers of units (cells) involved (i.e., no birth-and-death) and no interunit differentiation, then the number of cells belonging to each cell type in the system must be conserved. This is an important technical condition.

In this short space I can only sketch the treatment, and that only for the simplest cases, of morphogenetic movement in two dimensions. Let us imagine a square lattice, on which a family of cell types is to be distributed. A pattern is, as usual, generated by mapping this family $S = \{A, B, \ldots, R\}$ of cell types onto the lattice. The numbers N_A, N_B, $\ldots$, N_R, which represent the cardinalities of $F(A)$, $F(B)$, etc., and thus the number of cells of type A, type B, $\ldots$, which are initially present in the pattern, are now to be conserved.

We now heuristically introduce the two basic concepts considered by Steinberg—differential affinity and motility. Affinity is easy: two contiguous cells, say of type A and B respectively, define an *edge type* (AB) as their common boundary. If there are R cell types, there are R $(R + 1)/2$ possible contiguous pairs, and so the same number of edge types. To each edge we assign a non-negative number, which we write as λ_{AB}, λ_{AC}, $\ldots$, and which we think of as a "binding energy." By adding up all of these affinities for a given pattern, we generate a number associated globally with the pattern—a total binding energy.

Now we introduce a concept of local motility, embodied in a particular local state transition rule. Intuitively, we want a cell to be able to change positions with one of its neighbors, in such a way that (1) the local binding energy increases, and at the same time (2) the global binding energy also increases toward a maximum. It is not technically difficult to do this; I leave the detailed formulation as an exercise.

Detailed simulations show that the formalism just described is able to account for all experimental observations regarding cell sorting. Moreover, through perturbation of the parameters, we can also bring the formalism to bear on a variety of congenital malformations, and on such problems as invasiveness and metastasis in tumor growth.

As noted earlier, sorting is itself a general morphogenetic mechanism, playing a central role in the shaping of development. More than this, it imposes constraints on all the other morphogenetic mechanisms: They must be compatible or consistent with cell sorting, or else sorting will tear them apart.

Finally, the sorting algorithms are applicable far beyond their immediate interpretation as cell-sorting models. We have already mentioned the relation of sorting to phase separations of all kinds; this relation implies applications of sorting algorithms to a multitude of phenomena ranging from patching and capping of receptor sites to the distribution of species in an ecosystem. A more spectacular application, at the molecular level, arises from casting the problem of the three-dimensional folding of polypeptide chains as a sorting problem, subject to the constraint that the sorting units are not free but tied together in a linear chain.

Differential Birth-and-Death

Many morphogenetic processes involve the differential increase or proliferation of cellular populations and the corresponding diminution or loss of others. A good example of morphogenesis through differential cell death is the formation of fingers and toes: initially, there is a continuous tissue connecting all the digits, but ultimately the cells between the digits selectively die. Still more spectacular examples are found in insect and amphibian metamorphoses, in which the cells constituting the larval organs selectively die and are replaced by entirely new cell populations.

Perhaps the earliest formal treatment of such problems was that of Ulam, albeit in a highly idealized way. He considered the growth of patterns in two- or three-dimensional space, governed by recursive rules; new elements would be placed in a pattern, and old ones removed, according to the configuration of elements already present. To fix ideas, let us consider some of Ulam's examples, and then see how they may be represented in terms of networks.

1. Imagine the plane divided into squares (elements). Let us suppose that we are given a finite number of squares initially. We define a "growth rule" recursively as follows: the squares of the $(n+1)$ generation will be those of the nth generation, together with all those squares that neighbor exactly one square of the nth generation.

2. This example is the same as the preceding, except that we do not add a square in the $(n+1)$ generation if by doing so we would touch another square added at that generation.

3. This example is the same as example 1, except that we will assume our squares have a finite lifetime. For definiteness, we may suppose that all squares that have been in the pattern for two generations are removed in the third generation following their introduction. This allows squares to "grow back" into previously occupied regions of the plane.

Note that these examples are of an automata-theoretic character. To describe example 1 as a network, we shall suppose that the elements of our network are the squares in the plane. Each element may be in one of two states: 1, 0 (e.g., on and off, or alive and dead). Time is assumed quantized. The neighbors of an element n_i will be the four squares that share a side with it. We will assume that when an element is in the "on" state, it communicates an excitation of unit magnitude to its neighbors. The local state transition rule is then of the form

$$n_i(t + 1) = \Phi\left(n_i(t), \sum_{n_j \in U(n_i)} n_j(t)\right)$$
$$= 1 \text{ if } n_i(t) = 1$$
$$= 1 \text{ if } n_i(t) = 0 \text{ and } \sum_{n_j \in U(n_i)} n_j(t) = 1$$
$$= 0 \text{ otherwise.}$$

To describe example 2 as a network, we must enlarge the concept of neighborhood from that used in 1, to allow "inhibitory" signals to reach a square from "live" squares two steps away along the lattice. We then amend the local state transition rule so that a square becomes active at time $t+1$ if exactly one of its immediate neighbors is active at time t, and if it receives no inhibitory signal from any of its other neighbors (i.e., none of these cells is active at time $t+1$). I leave the precise statement of this rule as an exercise.

To describe example 3 as a network, we enlarge the state set of the elements by allowing several microstates to be associated with the active state. In the case being considered, two such microstates σ_1, σ_2 are necessary. (This σ is not related to that used previously for macrostate.) Intuitively, when a square is turned on at an instant t, it goes into the microstate σ_1. At the instant $t+1$, it goes into the microstate σ_2. At instant $t+2$, it goes into the inactive state 0. Both of the microstates σ_1, σ_2 have all the properties of the active state 1. Clearly, by adding more such microstates, we can allow the elements to have any finite lifetime. Once again, I leave the precise statement to the reader.

The so-called *tesselation automata,* or *cellular spaces,* were originally suggested by von Neumann (1966) in connection with his theory of self-reproducing automata. Consider two- or three-dimensional space partitioned in some regular way (say into squares or cubes). Suppose that each square contains a copy of some automaton, which may be in any one of a finite number of states (including "off"). As usual, the neighborhood of an automaton in such a tesselation consists of those with which it shares a common edge in the tesselation, and a state transition function is defined.

Von Neumann's original problem was the following. Suppose that we define an initial configuration of the tesselation, in which a finite number of the constituent automata are in active states, and all the other automata are "off." Can we define state transition rules such that, at the end of some definite time, the configuration of the tesselation will consist of two copies of the original active configuration, with everything else "off"? This question has been answered in the affirmative by a number of authors, and a large literature with strongly morphogenetic overtones has been elaborated within this framework. However, these problems are entirely network problems, interpretable as differential birth-and-death by allowing *birth* to mean the forcing of a neighboring inactive cell into an active state, and *death* to mean the opposite.

༈

CHAPTER 16

Order and Disorder in Biological Control Systems

Two basic characteristics of biological systems are the complex developmental processes which give rise to them, and the equally complex physiological processes which maintain them. These have been the subject of intensive study over the years because of their dramatic nature and their extreme practical importance. Not nearly so much attention, however, has been given to processes like senescence and mortality, which are in their way equally dramatic, equally important, and in a certain sense even more puzzling. Why should a physiological system, which is a collection of interacting homeostats designed precisely to buffer the organism against externally imposed stresses, not be capable of indefinite persistence? Why should organisms have rather well-defined lifespans, and why should these lifespans vary by factors of two, three, or more, between species otherwise taxonomically similar and obviously conforming to the same general *bauplan*?

Perhaps one reason that degenerative processes have been neglected in comparison with generative ones is that we intuitively expect all systems to fail eventually. Our own mechanical artifacts, with which we have had the greatest experience, do break down regularly; even with the most careful maintenance, machines and electronic devices malfunction and ultimately fail. We *expect* that the general trend of things will always go from "order" to "disorder." Biological phenomena, like development and physiology, are so striking precisely because they appear to violate this trend; that is why we must strive so hard to explain them. Senescence and mortality, on the other hand, appear to be fully consistent with the trend, and thus apparently (but only apparently) require no special explanations.

If development and physiology are respectively the processes through which order is generated and maintained, then we may regard senescence as in some sense a disordering. We are going to take the viewpoint that, in general, "disordering" in a system is manifested by the deviation of system behavior from some nominal *standard* of behavior. Furthermore, this standard against which system behavior is measured is, in the most general sense, a *model* of the system itself. This model or standard tells us how the system *ought* to behave; i.e., it generates expectations or predictions about the system. When actual system behavior coincides with that of the standard, this coincidence represents *order in the system.* When system behavior departs from that of the standard, we shall say that *the system is disordering* with respect to the standard; the degree of disorder is measured by the extent of the departure from standard behavior.

The reader may here object that in science, if a discrepancy arises between the actual behavior of a system and that predicted on the basis of some model, then it is the model which is to be rejected. That is, the traditional scientific method takes actual system behavior as the standard, and judges the model with respect to this standard. However, in biology, we freely invert this tradition. Thus, we speak of "errors" in the genetic code (e.g., mistranscriptions, mistranslations) whenever there is a discrepancy between, say, an actual generated polypeptide sequence and that expected on the basis of the code; we speak of neuronal "error" when a neuron fires even though its excitation is below threshold, or fails to fire when its excitation is above threshold. In all such cases, the "error," which is the discrepancy between actual behavior and nominal behavior, is imputed to the system and not to the standard or model. In all these cases, and many others which might be cited, we thus regard the *system* as disordering with respect to the standard.

Let us look at these notions in a limited but familiar context. A typical model of a biological process might take the form

$$dx/dt = F[x(t), \alpha(t)], \qquad (16.1)$$

where $x(t)$ is some appropriate state vector, and $\alpha(t)$ is a vector of controls or inputs or forcings. When integrated, these equations of motion specify trajectories of the form

$$x(t) \;=\; \Phi[x(\theta),\; \alpha(t)]. \qquad\qquad (16.2)$$

Thus, knowing an initial state, we can predict the system response to any particular control sequence $\alpha(t)$.

Now a control sequence $\alpha(t)$ generally arises because our given system is interacting with other systems; i.e., it is interacting with its environment. The existence of such interactions means that our system of interest is *open,* in the sense of thermodynamics. However, these equations of motion also specify precisely *how open* our system is assumed to be; i.e., it expresses absolute limitations on the character of system-environment interactions. That is, a model like 16.1 can tell us what happens if we make our system thermodynamically more close (i.e., if we abolish certain system-environment interactions characterized by 16.1), but it cannot in principle tell us what happens if we make our system more open.

Now any model 16.1 of an organic process is necessarily an *abstraction,* in the sense that features present in the system itself are neglected or omitted in the model. This is especially true with respect to potential interactive capabilities. In this sense, *our model of such a system must necessarily be more (thermodynamically) closed than is the system itself.* A basic corollary of this fact is the following: under very general conditions, the behavior of such a system will necessarily ultimately begin to deviate from that predicted on the basis of the model 16.1, and the magnitude of this deviation will tend to grow in time. In other words, if we take the model 16.1 as our standard of order, the behavior of the actual system will appear to become progressively disordered in time.

These considerations have some important implications, which we shall now briefly sketch:

1. What we have called "disordering" (i.e., the deviation of the behavior of a system from that expected on the basis of a given standard or model, which arises from the fact that the system is more open to interaction than is the standard) subsumes such concepts as *error, system failure,* and *emergence.* In all of these we find an unpredictable and apparently causeless departure from expected behavior, arising precisely because the interactions responsible for such behavior in the actual system have been abstracted out of our model, and hence are invisible in it.

2. Such "disordering" arises from the very nature of abstraction, and hence cannot be removed by simply passing to a more comprehensive model. Any model, however comprehensive, allows disordering with respect to it as long as it is an abstraction. Disordering of this kind is hence an essential fact of life. In this sense, it is reminiscent of the Gödelian limitations of axiom systems in arithmetics; in fact, the two kinds of limitations are closely related.

3. The employment of predictive models for the purpose of controlling a system (of which medical therapeutics is a specific instance) will generally result in a variety of unpredictable effects on system behavior ("side-effects") arising from the implementation of a control strategy. The appearance of such side-effects is in fact a familiar feature of attempts to control complex system behaviors. As we have seen, their appearance is a consequence of using an abstraction as the standard of system behavior, and cannot be eliminated by simply passing to a more comprehensive standard.

4. The disorderings we have described pertain to *any* behavior which is model based. This is true not only of manmade models, but also of the intrinsic models which biological systems generate and utilize to modify their own behavior. Indeed, much of the adaptive physiological behavior which has heretofore been considered exclusively in cybernetic terms (i.e., on *feedback* regulation through the generation of an "error signal") is best regarded as *feedforward* regulation based on such intrinsic predictive models. Such important physiological systems as the endocrine system and the immune system are almost entirely governed by such feedforward, model-based regulators. Accordingly, they are subject to exactly the same kinds of considerations as have been developed above. Ultimately, the systems they control will disorder relative to the control models, resulting in maladaptive responses and ultimate system failure. We have elsewhere (Rosen 1978d) proposed that such disordering is responsible for what we recognize as senescence in complex systems; we refer the reader to that work for fuller details and examples. We may note here, however, that one of the main features of system failure arising from such disorderings is in a sense that the failure is global and not local; i.e., the system as a whole can fail, even though each part of the system, considered in isolation, does not fail in any detectable way.

From these brief considerations, we can see that the "disordering" which occurs in biological systems can be vastly different from that characteristic of physical systems, which are related to increase of entropy via the Second Law of Thermodynamics. The disorderings we have considered above arise from the fact that a real system is more open to interaction than any abstract standard or model can be, and is thus more closely akin to the ideas of "wear-and-tear" which have often been invoked to account for system failures in biology and elsewhere. In physics, on the other hand, disordering arises from closing and isolating a system. Since thermodynamic ideas have been so influential in establishing the trend of systems from "order" to "disorder" which we mentioned at the outset, we will conclude our discussion by relating our considerations to those developed for thermodynamic systems.

In thermodynamics, the standard systems are those which are closed and isolated (i.e., those which are exchanging neither matter nor energy with their environments). Such systems can be shown to relax toward a unique equilibrium state (thermodynamic equilibrium). In simple systems, like fluids, this state of thermodynamic equilibrium is one of spatial and temporal uniformity possessing no gradients of any kind. This state of thermodynamic equilibrium is chosen as the *standard of disorder;* hence any nonequilibrium state of such a system is necessarily ordered with respect to it. Moreover, the dynamical process of relaxation toward equilibrium in such a system, which must begin initially with an ordered state and asymptotically approach equilibrium, is one of *disordering.*

A certain thermodynamic state function, the *entropy,* always increases as such a closed, isolated system relaxes toward equilibrium and assumes its unique maximum on the equilibrium state itself. In fact, entropy is a Lyapunov function for the dynamics; indeed, the assertion that entropy is a Lyapunov function, together with the necessary linearity of this dynamics near equilibrium, are essentially sufficient to completely determine the near-equilibrium dynamics. This is how the rate equations of "irreversible thermodynamics" are derived (de Groot and Mazur 1963; Prigogine 1967).

For closed, isolated systems, then, we have three equivalent propositions:

1. Such systems autonomously relax toward a state of thermodynamic equilibrium.

2. The state of thermodynamic equilibrium is the standard of *disorder* for states; hence every nonequilibrium state is ordered with respect to it. Further, the relaxation toward equilibrium is necessarily a disordering of the system itself.

3. A certain specific state function (entropy) is a Lyapunov function for dynamics in such systems.

It is this last proposition which gives entropy its special role in thermodynamics, and which allows us to identify "disorder" in thermodynamic states with the value of the entropy on those states.

If we now take such a closed, isolated system and open it up to environmental interaction. several things happen. First of all, the new forcings imposed on the system by virtue of these interactions will obviously change the dynamics of the system. The entropy function, which was a Lyapunov function for the autonomous dynamics when the system was closed and isolated, will in general have no special relation to the new dynamics, and loses thereby all its original theoretical significance.

Second, because the dynamics has been changed, new stable attractors appear, and the system will now autonomously move toward the one in whose basin of attraction it happened to lie initially. From the thermodynamic standpoint, of course, such movement away from thermodynamic equilibrium represents an *ordering* of the system with respect to the thermodynamic standard of disorder. For this reason, there have been numerous attempts to regard biological morphogenesis, and other kinds of pattern generation, in this light, simply on the grounds that biological systems are ordered in some absolute sense.

On the other hand, from the standpoint of the new dynamics in the open system, the autonomous movement of the system toward an attractor is exactly analogous to the relaxation of a closed isolated system toward equilibrium (which is the only available attractor for this dynamics). If we take the thermodynamic parlance seriously, the proposition 2 above says that the conditions of maximal disorder for an arbitrary dynamical system are determined by its attractor; these are what the systems autonomously tend toward, and what energy expenditure is required to keep them away from. That is, if some open system is chosen as standard, then that system necessarily disorders with respect to itself as it approaches its attractors. Moreover, with respect to this new choice of standard, *any* change in the character of system-environment interactions (whether their effect is to open the system further, or to close the system) will cause a deviation from this standard, and hence by definition will *order* the system.

Thus, from our standpoint, the thermodynamic characterization of order and disorder is inappropriate and unnatural in several ways. First, it chooses the dynamics of closed isolated systems as standard. It then chooses the attractors of such dynamics (thermodynamic equilibria), which are the states to which such dynamics autonomously tend, as standards of *disorder*. Finally, it measures disorder, according to these standards, by the values of entropy, which happens to be a Lyapunov function for the standard dynamics, hence identifying entropy with disorder.

On the other hand, we have argued that there is nothing special about closed, isolated systems; indeed, from the standpoint of biology, they are most unnatural systems to choose for standards. We have seen that any dynamics can be chosen as standard, and once we have left the thermodynamic realm, it is much more natural to regard this standard as establishing the norm of *order* and not disorder. From our viewpoint, thermodynamics represents a very special case of our treatment, *if we interchange the terms "order" and "disorder."* Indeed, the thermodynamic usage of these terms has no particular justification, except that of historical tradition.

As we have seen, the concept of entropy loses all significance once we depart from the choice of closed, isolated systems as standards. We can retain the term for open systems, but only at the cost of inventing an increasingly unwieldy formalism of "entropy fluxes," which are attempts to create new Lyapunov functions for the open system dynamics. If the system is open enough, this whole approach fails; the problems associated with open systems are dynamical problems and not thermodynamic ones.

Instead of entropy, the treatment sketched above gives as a measure of disorder the discrepancy between the actual behavior of the system and that of the standard or model which we have chosen to characterize order (or any numerical function of this discrepancy). Indeed, thermodynamic entropy is a very special case of this, in which the standard is the equilibrium of the system when closed and isolated, and with the terms "order" and "disorder" interchanged.

The discussion we have provided is but one small part of the larger problem of establishing the relation between a real system (physical, biological, or social) and the models we can make of that system (see Rosen 1978b). This is not only the basic problem of theoretical science, it is also central in understanding the behavior of systems which contain models, either of themselves or of their environments, and utilize these models for purposes of control. The existence of such internal models of self and/or *environment* is one of the essential differences between biological systems and nonbiological ones, but their presence has been essentially neglected in our reductionistic haste.

CHAPTER 17

What Does It Take to Make an Organism?

My late colleague and friend James F. Danielli liked to say that the progress of human culture depends on the capacity to move from an "age of analysis" to an "age of synthesis" (Danielli 1974:1). He viewed such a progression, in our own day, as an immediate practical urgency. Even thirty years ago, he was arguing that the capacity to synthesize life provided perhaps the only road to our continued survival as a species; that our deep problems of pollution, resource exhaustion, and ultimate extinction require giving up the machine-based technologies that create these problems and going to biological modes of production built around artificial organisms of our own design.

Be that as it may, the difficulty is that analysis and synthesis are not mere inverses of one another. We are very good at (at least some) modes of analysis, but we are not very good at all at synthesis. Indeed, most of our modes of analysis tend to be *inherently* irreversible: "Not all the King's horses, nor all the King's men . . ."

In biology, we continue to believe that the synthesis or fabrication of life involves nothing more than reassembling a class of analytic fragments, and indeed a very special class of analytic fragments at that. Above all, there is no theory to guide such efforts; in the last analysis, the most crucial ingredient in these attempts is chance, luck. And so far, we have not been very lucky.

To attach synthetic significance to analytic fragments and, still more, to believe that analytic knowledge (i.e., a knowledge of how something works, its *physiology*) can tell us something about its creation, its *ontogenesis,* is an article of faith inherited from the machine metaphor (see later section on machines and organisms, and chapter 20). It is a most unfortunate legacy: in too many ways, it leads us in exactly the wrong direction. It identifies analysis with synthesis, replacing the latter by the former. What we must do, rather, is to separate them again. The process of doing this, however, takes us immediately outside the machine metaphor itself, and everything it rests on.

That process, of distinguishing between analysis and synthesis in a biological context, is mainly what the following discussion is about. I make a basic distinction between a system N, and what we may best symbolize as $\exists N$. Fabricating N means entailing $\exists N$ from something. And as we shall indicate, formulating the circumstances under which this can be done turns out to exactly parallel what *already* goes on in organisms.

In fact, the concept symbolized by $\exists N$ does not generally belong to the epistemology of N at all. Insofar as contemporary science is wrapped up entirely in the epistemology of N, that symbol is scientifically meaningless; it has no epistemological significance. It belongs rather to what philosophers call *ontology*. It involves its own kind of causal structure, different from that embodied in N itself—a structure that in the philosophical literature is termed *immanent causation*.

I have come to believe that a theory of fabrication, and in particular a theory of fabrication of organisms, leads us directly to ontological questions of this type. These are questions that have been excluded from science, since at least the time of Newton, and left entirely in the hands of metaphysicians. Their language is therefore the only one presently available for articulating them; I hope to show, however, that there is nothing unscientific about the questions themselves. Quite the contrary, in fact.

Why the Problem Is Hard

Any problem becomes hard if one is inadequately equipped, or if the tools at hand are not appropriate to shaping the solution. The present general belief that the only alternative to vitalism is to try to anchor biology inside contemporary physics asserts that contemporary physics exhausts the resources of material reality, and hence that biology (or any other science) is only a special case of this ultimate generality. This is essentially *reductionism*.

Contemporary physics, the physics we know and that is found in the books and journals we read today, rests on assumptions that limit it profoundly. Accordingly, it is a very special science, applicable only to very special material systems; it is inherently inadequate to accommodate the phenomena at the heart of biology. No amount of sophistication within these limitations can compensate for the limitations themselves.

This is not to assert that biological phenomena are inherently unphysical. It is only to say that organic phenomena pose a challenge for contemporary physics at least as great as radioactivity did in 1850, or atomic spectra and chemical bonding did in 1900. As in those situations, to cope with such problems involves new physics, because the inherited physics is too narrow, too limited to meet these challenges directly. The hard part is here.

In this section, I will content myself with a few general remarks, mainly of a historical nature, which will supply the context and perspective for later developments.

The physics, the science, that we know today (see my *Anticipatory Systems*) is, in conceptual terms, largely where Newton left it 300 years ago. The revolutions that have periodically transformed it, such as relativity and quantum mechanics in the present century, have not touched these conceptual foundations; they pertain rather to a more sophisticated view of the relation between observation and the inherited framework of states and state transitions.

Before Newton, what passed for science was part of philosophy (natural philosophy). In many ways, this pre-Newtonian science was much richer than what has replaced it; Newton represented a radical *simplification* of science, a simplification that at the same time asserted *no loss of generality*. It was the simplification, as much as the unprecedented practical success of Newtonian ideas in the context of celestial mechanics, that recommended it, as summed up in the immortal words of Laplace: "*Je n'ai pas besoin de cet hypothèse.*"

Nevertheless, pre-Newtonian science *was* more complicated than the Newtonian, just as primitive languages are often more complicated than what replaces them. How did this pre-Newtonian science proceed? What were its modes of analysis? Briefly, using modern terminology, it asserted something like

CONCRETE SYSTEM = EXISTENCE + ESSENCE,

which seems unbearably quaint and archaic to our modern minds. *Existence* is the province of ontology, and it connotes roughly how the concrete system got to be what it is, from whatever it was before. It is thus concerned with evolution or development—with *creation.* From this perspective, an understanding of the ontology, the existence of a system, is an essential and independent part of understanding the system.

Essence, on the other hand, is roughly the province of epistemology. In practical terms, this part deals with how the system behaves—i.e., with its physiology rather than with its evolution or development. This could be analyzed further, into something like

ESSENCE = GENERAL + PARTICULARS,

an analysis that goes back at least to Plato; in our own time, it appears in, for example, the concept of gestalt, and even here in the concept of general system.

This provides the basis for comparing or classifying different systems: two systems are related through the generalities they share, and distinguished through their particularities. Such a partition is still at the heart of a science such as biology, where we feel that individual, particular living organisms are instances of *life* (the general), and that we can learn about this life by studying the particularities of living.

It may be helpful to regard the GENERAL as analogous to a *noun* in language, and PARTICULARS as analogous to qualifying adjectives.

The idea of *essence* was associated with the modes or categories of causality that originated with Aristotle, and they are well known (see Rosen 1985a, 1985d for how they survive in the Newtonian scheme of things). On the other hand, the notion of *existence* (which included creation) revolved around a different kind of causation, *immanent* causation, the entailment of existence.

The crucial thing about these pre-Newtonian modes of analysis is that, in them, all of the analytic categories are *independent.* Specifically, this means (1) existences need not in principle entail anything about essences, or conversely; and (2) generalities need not entail anything about particulars, or conversely.

In these terms, the Newtonian simplification amounted to *a denial of the independence of these modes of analysis.* In practice, this amounts to two hypotheses: First,

EPISTEMOLOGY ENTAILS ONTOLOGY.

This is a very powerful hypothesis because it asserts in effect that the origins or creation of systems is subsumed under the study of the operation of systems. In other words, it mandates that ontology is redundant, not a separate or independent study.

Next, in its way an equally powerful hypothesis,

PARTICULARS ENTAIL GENERALITIES.

Putting these two hypotheses together, we may write,

PARTICULARS ENTAIL EVERYTHING.

And that is where we still are today.

Seen in this light, the Newtonian simplification amounts to the imposition of *constraints* on the natural philosophy that came before. By the term *constraint* I mean an assertion of dependence between things which were initially independent. The effect of any such constraint is to single out, from the original universe of discourse, *that special class of things that happen to satisfy the constraints.* But the principal hypothesis, the one on which contemporary physics ultimately rests, and which does all the damage, is this: *The special class that we have singled out by imposing the above constraints is, in fact, the whole class.*

From this admittedly thumbnail sketch, we can begin to see how much metaphysics is actually involved in the Newtonian simplification. So much for his "*Hypothesis non fingo.*"

Once we are in this constrained world, we cannot get out again, at least not from inside. All we can do is apply still more constraints, entirely within that already-constrained framework. And if this is *not* the whole world, if there are aspects of material reality that sit outside its grasp, then we cannot reach them from inside; if we insist on staying inside, we are in serious trouble.

I contend that organisms take us outside this special world, into a realm where particulars do not necessarily entail generalities and where knowledge of how a system works does not entail how it is created. Since our problem is with *fabrication* (i.e., with ontology), we simply cannot solve it by epistemological means alone in a world where ontology and epistemology are different.

That, however, is precisely what we have been trying to do. And that is why the problem is hard.

Relational Biology

Before pursuing the implications of the foregoing, I shall add another thread to the tapestry.

Rashevsky's term *relational biology* (see chapter 15) described a radically different approach to biology and to organism, one that ultimately laid bare the special character of contemporary physics and the attendant inadequacies I have sketched. Before contemplating the essence of what he did, let us first see how he came to do it; it is instructive and pertinent.

Rashevsky was himself trained in theoretical physics, and he believed in it with his whole soul. He began concerning himself with biological problems because they seemed to stand aside from physical comprehension, and he believed that no material phenomenon should do that. In that sense, he was a reductionist, at least as much as any contemporary molecular biologist.

For three decades, Rashevsky threw his formidable energies into creating a "mathematical biology" that would, as he put it, do for empirical biology what mathematical physics does for experimental physics. He proceeded entirely along what we now regard as conventional lines (though they were anything but that in his own day); he isolated specific biological processes, such as cell division or nerve conduction, and brought physical principles to bear on them in the form of explicit models. He was very good at making such models. For instance, he discovered the morphogenetic capabilities of reaction-diffusion systems, the subject of so much current infatuation, at least a decade before Turing ever heard of them, as a corollary of his work on cytokinesis. It was Rashevsky who invented the idea of the neural network, and thereby turned what started as a study of peripheral nerve excitation into a theory of the brain itself.

Rashevsky was animated by the conviction that organism would emerge, as if of itself, as the concatenation of sufficiently many such models. But after a while, he had to admit that this did not seem to be happening. In fact, he had to admit that the models seemed somehow to be *estranging* him from the organism; the more its different properties were captured in such models, the more the organism itself seemed to retreat from view. (Today, for analogous reasons, the organism is likewise disappearing in a welter of molecules.)

On the other hand, Rashevsky had embarked on the strategy of modeling as a way of understanding organism; it was organism that was the object of these studies, not its individual aspects in themselves. Putting the two together, he concluded that he would have to radically revise his strategy. The result was relational biology.

Relational biology can be thought of as the exact inverse of reductionistic ideas. The essence of reductionism is, in a sense, to keep the matter of which an organism is made and throw away the organization, believing that the latter can be effectively recaptured from the former. Rashevsky's relational biology sought rather to keep the organization and throw away the matter; to treat the organization itself as a *thing*, and recapture specific material aspects through a process of *realization*.

In this view, then, an *organism* is a material system that realizes a certain kind of relational structure, whatever the particular material basis of that realization may be. The trick, of course, is to find or posit that relational structure; this is not an empirical or experimental problem in any conventional sense. Rashevsky himself got only as far as exploring what could be said just from a knowledge or presumption that such a common relational structure *exists*, without necessarily knowing what it is. This was the basis for what he called the study of "abstract biologies"; it turns out that one can say quite a lot, just on this basis.

Rashevsky's own forays into relational biology all revolved around the idea of biological *functions*, and the relations between them, totally independent of the explicit material machinery for executing them, or manifesting them, or carrying them out. All of these words (e.g., *function*, *organization*) are, of course, totally alien to the physics of Rashevsky's own experience—a physics that regards them as mere epiphenomena when it considers them at all. On the other hand, Rashevsky's relational graphs can be thought of as constituting molecules of functions, retaining the modularity we are used to, but now transported to an entirely new realm.

I do not think that Rashevsky was fully aware of how radical his approach actually was. I do not think he was aware that he was doing new physics, or that his relational ideas are literally living violations of the constraints underlying the Newtonian simplifications I addressed previously. But before returning to these general matters, let us look at one class of relational models in a little more detail.

The (M, R)-Systems

I devised a class of relational cell models called (M, R)-systems (*M* for metabolism, *R* for repair). The idea behind these systems was to characterize the minimal organization a material system would have to manifest or realize to justify calling it a *cell*. It seemed natural to do this; the cell is important in biology, elevated into a basic reductionistic unit by the cell theory. Yet so many different kinds of things, from bacteria to neurons, are called cells that, if this terminology is to be meaningful at all, it cannot have an exclusively material basis.

It seemed to me (and still does) that one would not call a material structure a cell unless its activities could be partitioned into two classes, reflecting the morphological partition between nucleus (genome) and cytoplasm (phenome), and the corresponding functional partition between what goes on in cytoplasm (the M of the system) and what goes on in nucleus (the R).

In relational terms, this requirement (together with a few biological simplicities and mathematical compactions) boils down to a refreshingly simple diagram as shown in figure 17.1, where A and B are just sets, f is a mapping, and $H(A, B)$ is a *set of mappings* from A to B. It is not too misleading to think of f as a kind of abstract enzyme, which converts substrates $a \to A$ into products $f(a) \to B$, and to think of Φ as a process that converts these products into new copies of such an abstract enzyme. Chasing an element through figure 17.1 provides a time frame. (For fuller details of the arguments involved, see Rosen 1972.)

There is yet no *replication* in sight—nothing to make copies of the map Φ. We could, of course, simply augment the diagram with another mapping that would do it, but this would avail little except to lead into an incipient infinite regress. Thus I was edified to discover that, under

$$A \xrightarrow{\;f\;} B \xrightarrow{\;\Phi\;} H(A, B)$$

FIGURE 17.1

stringent but not prohibitively strong conditions, such replication essentially comes along for free, requiring nothing else but what is already in the diagram.

Let us sketch the simplest way this can come about (although it is not the only way). Quite generally, if X and Y are sets, and $H(X, Y)$ is a set of mappings from X to Y, I can always regard the elements $x \rightarrow X$ as operators $\hat{x}$ on $H(X, Y)$, by the simple device of defining

$$\hat{x}(f) = f(x).$$

For obvious reasons, $\hat{x}$ is called the evaluation map associated with $x \rightarrow X$. Then,

$$\hat{x} \rightarrow H[H(X, Y), Y].$$

If this evaluation map has an *inverse* $\hat{x}^{-1}$, then

$$\hat{x}^{-1} \rightarrow H[Y, H(X, Y)].$$

When will it have an inverse? If and only if

$$\hat{x}(f_1) = \hat{x}(f_2) \text{ implies } f_1 = f_2,$$

or

$$f_1(x) = f_2(x) \text{ implies } f_1 = f_2.$$

But this is a condition on $H(X, Y)$; if two maps in it agree at x, they must agree everywhere on X.

As stated, this is true for any sets X and Y. In particular, let us put

$$X = B$$
$$Y = H(A, B),$$

where A and B are the sets in the (M, R)-system (figure 17.1). It is easy to see that, if an evaluation map in this case is invertible, it is precisely a replication map—i.e., a map into the set

$$H[B, \ H(A, \ B)]$$

to which Φ belongs. And it is also easy to verify that it must have all the biologically important properties of replication.

Thus there is no need for an infinite regress. Under the appropriate conditions, metabolism and repair already entail replication. And this follows from organization alone, independent of any details of how the organization is realized. To my knowledge, this result still remains the only one in which replication is entailed by something—where it does not have to be posited independently. And this is the main reason I have remained interested in them.

From this point of view, then, a *cell* is (at least) a material structure that realizes an (M, R)-system, under the condition that at least one of the appropriate inverse evaluation maps exists. *Making a cell means constructing such a realization.* Conversely, I see no grounds for refusing to call such a realization an autonomous life form, whatever its material basis may be.

The (M, R)-systems have many other interesting properties. In the next section, these systems will arise automatically in a completely different context—indeed, a most surprising one—when we drop the constraints imposed by what I earlier called the Newtonian simplification.

Realization Problems

To tie all these diverse ideas together, we need one further concept, that of *realization.* The problem of "artificial life" is itself nothing but a realization problem. We want to produce material systems that manifest those general attributes that we associate uniquely with living organisms. In the quaint philosophical language introduced in the section called Why the Problem Is Hard, we want to fabricate a system that will manifest a particular essence. We want to build such a system, to *make it exist.*

In science as we know it, on the other hand, we always go the other way. We start from a system in the world (let us call it *N*), whose *existence* is therefore presumed, and we seek to *discover* its attributes, its essence. This activity culminates in the making of (formal) models, whose inferential structure is in congruence with causal structure in N. In fact, we concluded that everything that we can know about N, its entire epistemology, is wrapped up in these models and the relations between them.

So in passing from realization N to specific models or attributes [which for convenience we will abbreviate as $A(N)$], we habitually ignore ontology; we simply start from the presumption that there *is* an N (i.e., $\exists N$) such that $A(N)$; indeed, the problem here is specifically to characterize $A(N)$ so that the entire statement

$$\exists N \;\rightarrow\; A(N) \tag{17.1}$$

is true.

In the realization problem, however, we are *given* $A(N)$, and the problem is to make $\exists N$ true, or, better, to make the entire statement 17.1 true. Thus the problem in realization is with the ontological part of statement 17.1, not with the epistemological part.

It is here that the first Newtonian simplification

$$A(N) \;\rightarrow\; \exists N \tag{17.2}$$

simply fails. Indeed, it is the primary problem of ontology to find conditions that *can* entail a proposition such as $\exists N$ at all.

Indeed, it can be shown that the constrained world we have described in the Why the Problem Is Hard section simply admits no entailments such as statement 17.2 at all. It is a world in which we can find no *logical* reason for the existence of anything outside of it. Within this world, then, we cannot even meaningfully formulate, let alone solve, a realization problem. I shall touch on this in the next section when discussing the concept of the machine.

Thus we begin to glimpse the kind of world in which realization problems can be formulated—a world in which ontology is possible. More specifically, we need a world with "enough" entailment so that a proposition such as $\exists N$ can itself be entailed from something. But this is still not enough; as it stands, it is purely formal, and it refers only to the proposition $\exists N$. We need to refer not just to $\exists N$, but to N itself. In short, we need

IF ∃ N CAN BE ENTAILED, THEN N ITSELF CAN BE ENTAILED.

This, in brief, is where *immanent* causation enters the picture.

Furthermore, if N can be entailed, then there is something (call it R) that entails it. We can thus write

$$R \Rightarrow N,$$

or, better,

$$R \Rightarrow (X \Rightarrow N).$$

But this is essentially the description of an (M, R)-system.

Thus we can conclude: The solution of *any* realization problem automatically generates an (M, R)-system! Notice how, in the very formulation of such realization problems, ontology and epistemology must be initially separated, and then intertwined. It becomes all the more interesting when it is precisely an (M, R)-system that we want to realize.

More specifically, from this point of view, an (M, R)-system represents *a model of a realization process.* Consequently, realizing an (M, R)-system involves an iteration of the very process that generates it. We see here a reflexivity that cannot even be formulated in the world of the Why the Problem Is Hard section.

Let us note further that all we established in the preceding argument is ∃ R. We know from that argument *nothing else about R*—only that it exists, and that it entails N. These are not enough to define R as a system in itself, with its own attributes, its own models, its own descriptions. But we *do* know that, whatever R itself may be, R itself can be entailed, because we just succeeded in entailing ∃ R.

This is where the magic of *replication* comes into the picture. In an (M, R)-system with replication, as described in the preceding section, we have enough entailment to close the realization process up on itself. This kind of realizational closure is close to the very essence of organism. It is, quite literally, a complex business; it involves a great deal more entailment than we are allowed in the constrained world of the Why the Problem Is Hard section. In a sense, that constrained world admits a *minimal* amount of (causal) entailment; the world of organism requires much more than that.

Machines and Organisms

The relation between biology and mechanism has always been an uneasy one, but it has been extremely influential nevertheless, never more so than today. Since it constitutes the essence of reductionism and has shaped most serious attempts at both the analysis and synthesis of organisms, including the one chronicled here, I must discuss it briefly.

The machine metaphor was first proposed by Descartes in the early seventeenth century. It is reported that, as a young man, Descartes was much impressed by some lifelike hydraulic automata. With characteristic audacity, he later concluded from these simulacra that life itself was machinelike (rather than that machines could be made to appear lifelike, which was all he was really entitled to conclude)—despite the fact that he had only a feeble notion of what an organism was and, as it turned out, even less of what a machine was. We must remember that, at this time, Newton was still a generation or two in the future.

Indeed, considered as material systems, machines are still hard to characterize in intrinsic (i.e., epistemological) terms, independent of their origin or history. Yet the idea that the machine, or mechanism, is the general, and biology only a particular, remains a compelling and fascinating one.

Nevertheless, sooner or later, we must come to grips with the question (to paraphrase Schrödinger), What is machine? Let us pursue this a bit.

One kind of answer to this question, at most half a century old now, was distilled from a still potent brew or amalgam of neural networks, Boolean algebra, computation, and studies in the foundations of mathematics. This is wrapped up in the notion of the mathematical machine, by which everyone nowadays understands "Turing machine." The Turing machines are symbol manipulators, string processors, whatever one wishes to call them. They are coextensive with the idea of algorithm; what they do is called *effective*. The identification of effectiveness with algorithms, and thence with what Turing machines do, is the thrust of Church's Thesis (see chapter 10).

The manipulation of meaningless symbols by fixed external rules is, it should be noted, exactly analogous to the Newtonian view of material nature, expressed in terms of manipulation of configurations of structureless particles by impressed forces. As we shall see, the two are indeed, at root, exactly the same.

The mathematical interest in these mathematical machines is based in concerns with axiom systems and their consistency. One approach addressing these concerns, initiated by David Hilbert, came to be called *formalism*. In this approach, mathematics was viewed as a game of pattern generation or symbol manipulation, governed by a finite number of fixed rules. Hilbert asserted that all of mathematics could be reformulated in these terms, *without loss of truth,* but at the cost of complete loss of *meaning.*

This last is the crux, and it must be described in a bit more detail. Mathematics, like any language, has two aspects, *syntactic* and *semantic.* The first pertains to the internal rules governing the structure of the language itself—its internal grammar. The second pertains to what the language describes or is about—to external referents, to meanings. Hilbert believed that all the foundational troubles within mathematics were arising from the semantic aspects—the idea that mathematics had to be *about* something. He believed that the semantic features of mathematics could be completely replaced by more syntax, without loss of truth; the actual accomplishment of this replacement of semantics by syntax was the formalist program.

The goal of this program, then, was a language of pure syntax, without semantics at all. Put another way, the idea was to strip mathematics of all referents whatever, and to replace them by equivalent, but purely syntactic, internal rules.

A mathematical formalism (e.g., Euclidean geometry or group theory) that could thus be stripped of its semantics is called *formalizable.* Hilbert thought that every mathematical formalism was formalizable in this sense. However, the formalist program was wrecked by the Gödel Incompleteness Theorem, which showed that Number Theory is already not formalizable in this sense. In fact, Gödel (1931) showed that any attempt to formalize Number Theory, to replace its semantics by syntax, must lose almost every truth of Number Theory.

What has all this to do with machines and life? Well, the epistemology of any material system is bound up with the set or category of all of its models. A model is a formalism, an *inferential* structure, that can be brought into congruence with *causal* entailments in the system being modeled. The modeling process is thus one of supplying *external referents* to the formalism, in such a way that what goes on in the external referent (the system), and what goes on in the model, become coincident.

Now any mathematical formalism per se, be it a model of something or not, may be formalizable, or it may not be. If it is formalizable, then it can be replaced by an equivalent formalization—i.e., by a set of internal, purely syntactic rules. It can thus be replaced by a mathematical machine, a Turing machine. In a word, it is *simulable*. But simulability means that all external referents, and hence all semantics, have been internalized, in the form of equivalent syntactic rules—by hardware in a machine or simulator. From inside a formalization, then, there is no way to establish that *anything outside it exists*.

In the context of natural science, we can characterize a class of material systems by requiring that all of their models be simulable. Let us call such natural systems *mechanisms*. Church's Thesis then asserts, *mutatis mutandis,* that *every material system is a mechanism*. This is, of course, the essential underlying proposition of all of contemporary physics. From these ideas, it is easy to see that all the analogy between syntactic symbol manipulation and Newtonian mechanism lies very deep indeed.

For our purposes, the main thing to notice is that, in such a world, *there is no ontology* because there is no semantics, no external referent. This is the heart of the first Newtonian constraint or simplification mentioned in the Why the Problem Is Hard section. In such a world, ontology has been swallowed by epistemology, as semantics has been swallowed by syntax.

This kind of world is a very feeble one, as measured by the kinds of entailments possible in it. It also is a very special world: it is highly nongeneric for a formalism to be formalizable. And yet this is the world that, according to the machine metaphor, must be still further specialized to get to biology.

Indeed, it can be shown that the Newtonian constraints *characterize* this world of mechanisms. That is, they already mandate the simulability of all models.

Loosening the Newtonian constraints thus means asserting the *existence* of material systems with nonsimulable models. In such a world, semantics cannot be reduced to or subsumed under syntax; ontology has been decoupled from epistemology. Only then can we talk meaningfully about the existence of external referents, which manifest themselves precisely in those semantic aspects that cannot be formalized.

It is also only in such a world, where there are autonomous ontological aspects, that we can meaningfully talk about immanent causality, and hence about fabrication. Hence we see that it is no accident that contemporary science offers no *theory* of fabrication; within it, we can only go from what is already special to something more special still; from something that is already constrained to what is still more constrained. But this is the very essence of simulation.

On these grounds, we can see that the fabrication of something (e.g., an organism) is a vastly different thing than the simulation of its behaviors. The pursuit of the latter represents the ancient tradition that used to be called biomimesis, the imitation of life (see chapter 7). The idea was that by serially endowing a machine with more and more of the simulacra of life, we would cross a threshold beyond which the machine would *become* an organism. The same reasoning is embodied in the artificial intelligence of today, and it is articulated in Turing's Test. This activity is a sophisticated kind of curve-fitting, akin to the assertion that since a given curve can be *approximated* by a polynomial, it must *be* a polynomial.

In conclusion, *any material realization of the (M, R)-systems must have noncomputable models.* Hence they cannot be mechanisms, in the sense we have been using the term, because the (M, R)-systems are too rich in entailment to be accommodated within the world of mechanism. This kind of result should not be unexpected, given the relation between the (M, R)-systems and the nature of realizations, something that involves external referents (and hence inherent nonformalizability) in an essential way.

Thus we have at least part of an answer to the question with which we started—What does it take to build an organism? It takes a lot more than we presently have. That is why the problem is so hard, but also why it is so instructive.

Part V

⌒

ON BIOLOGY AND TECHNOLOGY

T HE CHAPTERS in this part are of a different character from those preceding. They bear not so much on what we can learn about biology from other disciplines as on what we can learn about other disciplines from an understanding of biological modes of organization. Most particularly, they bear on technologies—how to solve problems. Here I shall use *technology* in the broadest sense, to include problems of an environmental and social nature, not just the fabrication of better mechanical devices, and to connote the execution of functions.

I have long believed, and argued, that biology provides us with a vast encyclopedia about how to solve complex problems, and also about how not to solve them. Indeed, biological evolution is nothing if not this, but its method of solution (natural selection) is, by human standards, profligate, wasteful, and cruel. Nevertheless, the solutions themselves are of the greatest elegance and beauty, utterly opposite to the discordances and mortal conflicts that created them. We cannot use Nature's methods, but we can (and, I believe, we must) use Nature's solutions.

I have also long believed that there are many deep homologies between social modes of organization and biological ones that make it possible to learn deep things about each by studying the other. I believe the situation here is very much akin to the Hamiltonian mechano-optical analogy that I touched on in chapter 14, an analogy that enabled us to learn new and profound things about optics while studying mechanics, and vice versa (while having nothing to do with reducing the one to the other). The thread that weaves such disparate subjects together is rather of a mathematical character, a congruence between their distinct entailment modes—common *models* that are diversely realized. In this case, the models are relational, and they are complex.

The common relational models that bridge biology and the technologies allow us, in principle, to separate the fruits of selection without needing to emulate its methods. They provide a Rosetta stone that allows us to utilize the billions of years of biological experience contained in Nature's encyclopedia, and to realize them in our own ways, applied to our own problems.

These matters were all resolutely, although with great reluctance, excluded from *Life Itself.* However, they played an integral role in the development of the lines of argument detailed therein. For instance, the idea of (biological) *function* developed therein (in which a subsystem is described in terms of what it entails, rather than exclusively in terms of what entails it) has an indelible technological slant, which I exploit as a point of departure in the chapters of this part. I make many uses of this notion, even though it is dismissed by reductionistic biology as merely a vulgar anthropomorphism. It should be noted that this concept of function exists even in contemporary mechanical physics; it is closely related to the distinction between inertial and gravitational aspects of matter described in part I (see specifically chapter 1).

A metaphor I use to motivate the study of this biological encyclopedia in technological contexts is that of the *chimera.* In biology, this term connotes a single organism possessing more than the usual number of parents—e.g., whose cells arise from genetically diverse sources. The chimera is in fact a point of departure from biology into technological considerations, and this in many ways. Our civilization has become replete with man-machine chimeras, and even machine-machine chimeras, which manifest emergent functions their constituents do not possess. Social structures, and even ecosystems, are chimerical in this sense. Even such things as activated complexes in biochemistry can be regarded as chimeras. Yet they have been little studied, being looked upon in biology as mere curiosities.

However, the mysterious interplay between genotype and phenotype is deeply probed by chimera. And the notion of *function* is central. One aspect is that the interplay of function in chimeras is an inherently *cooperative* notion, not a competitive one. Indeed, one of the deepest lessons of biology is that such cooperation is selected for; indeed, that life would be impossible without it; and hence that complex organizational problems can be solved via cooperation and not by power and competition.

Actually, this is an old idea of mine. In 1975, I was invited to participate in a meeting entitled Adaptive Economics, despite my protests that I knew nothing about economics. Clearly, the organizers were of the opinion that *adaptive* is universally good, a word impossible to use pejoratively, and what was wrong with our economic system was, in some sense, its failure to be sufficiently adaptive. Equally clearly, they wanted me only to provide some biological examples of adaptation, to lend indirect support to this view. I thought I could easily provide a catalog of such, and set out to write a paper in this vein. However, I ultimately found myself writing something quite different. The lesson of biology turned out to be that adaptiveness is not universally good; too much of it, in the wrong places, will tear cooperative structures apart. Indeed, it turns out that organism physiology is very careful in its apportionment of adaptivity; survival depends on it. This is perhaps not the lesson the organizers wanted me to deliver from biology, but it is the one that biology itself wanted—one small excerpt from its encyclopedia. (Although not explicitly developed in that paper, there are close ties to my development of model-based anticipatory controls, which were proceeding concurrently at that time.)

Another thread in all these works is my warning about the *side effects* that arise inevitably when attempts are made to control a complex system with simple controls. These side effects generically cascade into a devastating infinite regress. Biology, seen in this light, consists of illustrations of how such cascading side effects can be forestalled or avoided; the result is, inevitably, a system with relational properties very like my (M, R)-systems. Specifically, there must be a characteristic backward loop, relating a "next stage" in such a cascade with earlier stages—a future with a past. This, it should be noted, is the hallmark of an impredicativity—one of the characteristics of a complex system, and one of the main pillars on which *Life Itself* is built.

The idea of function is resisted in orthodox biology because it seems to carry with it a notion of design, and it seems necessary to expunge this at any cost. This is because *design* seems to presuppose a category of final causation, which in turn is confuted with teleology. Nevertheless, Kant (in his *Critique of Practical Reason*) was already likening organic life with art, and the lessons of life with craft. In chapter 20, which deals with human technology in terms of art and of craft, and with the role of the biological encyclopedia in furthering these endeavors, many of the individual threads just reviewed are interwoven into a single framework.

An early attempt to pursue biological correlates of technology, and the converse technological correlates of biology, was pursued under the general (though diffuse and ill-defined) rubric of *bionics*. Chapter 19 is a review of the history of this endeavor; it flourished for less than a brief decade (roughly 1960 to 1970). As we note, all that exists of it today is the field of artificial intelligence—and that in a vastly mutated form based entirely on a concoction of software, very different from what was initially envisioned. A renewed and concerted effort in this direction, an effort to truly read the encyclopedia that biology has left for us, is an urgent national, indeed international, priority, in the face of the burgeoning problems faced by each of us as individuals, and by all of us as a species. Spending billions of dollars on a human genome mapping project, while ignoring the technological correlates of biological organization that bionics tried to address, is an egregious mistake—the very kind of mistake that leads organisms to extinction.

Chapters 21 and 22 deal with an approach to complex systems from the direction of dynamics. This direction I also reluctantly excluded from *Life Itself*, but it is of great importance, especially when combined with what is presented therein. What is most interesting is its inherent semantic, or informational, flavor, expressed in terms of, for example, activations/inhibitions and agonisms/antagonisms. Here, impredicativities and unformalizability appear in the guise of non-exactness of differential forms. And, of course, most differential forms are not exact.

In some ways, I regard the chapters in this part as the main thrust of this entire volume. I am always asked by experimentalists why I do not propose explicit experiments for them to perform, and subject my approaches to verification at their hands. I do not do so because, in my view, the basic questions of biology are not empirical questions at all, but, rather, conceptual ones; I tried to indicate this viewpoint in *Life Itself*. But the chapters in this part, I hope, expound the true empirical correlates of biological theory. In the realm of art and craft, rather than in a traditional laboratory, will ample verification be found.

∾

CHAPTER 18

Some Lessons of Biology

Perhaps the first lesson to be learned from biology is that there are lessons to be learned from biology. Nowadays, biologists would be among the first to deny this elementary proposition as nothing more than a vulgar anthropomorphism. To say there are lessons to be learned is, in their eyes, to impute a design to be discerned. And according to Jacques Monod (1971), one of the major spokesmen of contemporary biology, the absolute denial of design (he calls it the "Postulate of Objectivity") is the cornerstone of science itself. Science lets us therefore learn *about* things, but not *from* them.

This in itself is a very seventeenth-century view, perhaps first conspicuously propounded by Descartes—hardly contemporary. But Descartes was wrong in this regard. In fact, the perceived dichotomy between design and objectivity is an utterly false one. There are limitless domains that fall into neither category, and one of them is biology. Or, to put it into another (but still ancient) language: one can deny mechanism without thereby espousing vitalism, and vice versa. That is the first lesson of biology, from which all the others follow.

A humble biological situation illustrates some of these assertions. There is a family of crabs, collectively called hermit crabs, that are common on the coasts of Europe and North America. Unlike other kinds of crabs, the hermit crabs do not possess a rigid protective exoskeleton. That is a fact. Here is another fact: These crabs sequester abandoned snail shells, insert their soft abdomens into them, and proceed to carry them around. When they begin to outgrow an adopted shell, they seek another larger one into which they fit. And still another fact: hermit crabs collect sea anemones, to grow on the shells they have adopted.

I have two questions about these creatures. The first is primarily a structural one: How shall we account for the single system consisting of a hermit crab, plus its adopted shell, plus the anemones? As a single entity, it is an extremely composite, heterogeneous structure—what biologists often term a *chimera* (see also chapter 21). Its parts clearly have entirely different origins: the crab itself is arthropod, the shell comes from a completely different phylum (mollusk), while the anemones are from yet a third phylum (echinoderm). Yet the entire assemblage, considered as a unity, has its own form, or *phenotype,* presumably (at least, according to people such as Monod) coded for by a corresponding *genotype.* Thus our first question is really, What do conventional categories of genotype and phenotype mean when we are dealing with a chimerical creature like the hermit crab? (if indeed they mean anything at all).

The second question relates our hermit crab to the vast area we call *technology.* It seems very natural, and entirely harmless, to regard an adopted shell as performing a *function* for the crab, one that the crab is unable to perform autonomously, for itself—likewise the collection of anemones by this already chimeric entity. All this gives the appearance of *art.* My colleagues insist quite vehemently that concepts such as art must not be introduced here; there is only *program* of something somewhere. To even think of the adopted shell in functional (i.e., technological) terms is a violation of every canon of scientific objectivity.

The essence of chimera is summed up in the familiar American motto *E Pluribus Unum*: unity out of plurality. However, the entire thrust of biology over the past century, and of science itself since at least the seventeenth century, is rather to seek diversity out of unity. In biology, this is called *differentiation.* The idea that everything is differentiation has led biology to reject the concept of function, which is a perfectly rigorous notion, and replace it by a mystical concept of program. This replacement is, ironically, supposed to mechanize biology; to make it truly scientific. Instead, it has not only deformed biology almost out of all recognition but has made its lessons, particularly on the functional or technological side, unnecessarily difficult (if not impossible) to decipher.

In fact, we live in a highly chimeric world. Any ecosystem, for instance, is a chimera in the strictest biological sense; the functional distinction between, for example, predator and prey in such a system has nothing to do with differentiation or development as normally understood. Man-machine chimeras, and, increasingly, machine-machine chimeras, are becoming common in our world, and the thrust of our activity here is to make them still more common, and to make them better.

The biological concept of chimera demonstrates, on the one hand, the inadequacy of identifying organisms and their behaviors with mechanical programmability in the name of objectivity, and on the other hand (indeed, as a corollary), the invocation of rigorous concepts of functionality as crucial to biology itself, and as a bridge to technology. This concept makes old categories utterly obsolete and makes it impossible to sustain the prevailing notion that *science* and *mechanism* are synonymous.

So let us begin with the basic dualism of biology, that between phenotype and genotype of a given organism (see discussions in chapters 1, 3, and 6). This dualism goes back to Mendel, and in a different way to another early geneticist, August Weissmann.

Phenotype is a collection of properties and behaviors—what René Thom designates as forms. It pertains to what Weissmann called soma. In ordinary circumstances, it is a product of differentiation, arising through a process of morphogenesis. Mendel's contribution was to regard phenotype as *forced behavior.* What was doing the forcing came ultimately to be called genotype. By comparing a forced behavior of a system with a reference behavior *of the same system,* we generally get a (functional) description of what is doing the forcing—in this case, of genome. Mendel could not quite do this, but he did the next best thing: he compared two organisms from different generations but otherwise presumed identical except in "one gene." Phenotypic differences between them thus defined, or *measured,* something about genotype itself. Genotype is thus placed somewhere outside phenotype, but inside organism.

Now, what is the genotype of our hermit crab? Or indeed, of any chimeric system? To answer such questions, we must extend, or generalize, the terms *genotype* and *phenotype* beyond their customary realms. We can say that phenotype is forced by genotype in some basic fashion. The key concept then is in the word *forced.*

The concept of forcing goes back to an even more primitive dualism than genotype-phenotype, the dualism between *system* and *environment.* In its original mechanical manifestation, the definition of a system (e.g., as a set of particles we wish to pay attention to) automatically carried with it, and equally fixed, the environment of that system (e.g., all the other particles in the universe). But this bothered many people. Theologians did not like it because it banished free will, and, even worse, experimentalists did not like it because it made doing controls difficult. So it became possible to imagine the *same* system in a *multiplicity* of different environments. What this word *same* means will, in a moment, be tied to the biological idea of *genome.*

Since environment is the seat of (impressed) forces, the same system, in different environments, will behave differently. But it must still be the *same* system; hence we must decouple the identity of the system, which must be environment independent (i.e., objective), from its behaviors, which will of course vary from one environment to another. Or, stated another way, behaviors can be *forced* by what the environment happens to be; identity must be independent of this.

Note that the tacit forbidding of identity to depend on behavior, and hence on environment, has become the hallmark of scientific objectivity. One of its immediate corollaries is the injunction against functional descriptions, to which I called attention at the outset. A further corollary is a tacit identification of *all environments* with *all environments of which system identity is independent.*

To make a long story short, the concept of system behaviors is traditionally associated with *system states,* which change according to definite *laws* embodying the forcings impressed on the system by particular environments. These system laws typically contain additional quantities, usually called *parameters,* that mediate between system states and environmental forcings. I have discussed the causal correlates of such a picture, and in particular, the segregation of causal categories into separable structures in earlier writings (Rosen 1985a,d).

In such a picture, system identity must be associated with the parameters appearing in the system laws. These have to be presumed independent of everything—independent of system state and of system environment. In short, they may be regarded as determining the *species* of system we are dealing with. It should be clear that this concept of system identity, or *species,* represents a natural generalization of the biological concept of genome; and, accordingly, that the genotype-phenotype dualism, which arose initially out of biological considerations, is in fact of universal currency—a corollary of the system-environment dualism.

Although these considerations have made no mention of chimera, they apply to systems in general, chimeric or not. Accordingly, a chimeric system, like a hermit crab with its adopted shell and its polyps or medusae, will have behaviors of its own, just as any system does. These behaviors (let us call them phenotypes) will be expressed in terms of indigenous state variables and governed by system laws. In these laws will appear parameters, or genome, independent of both system and environment; the genome will mediate between them and convey on the system its characteristic and unvarying identity. This much is presumed true for any system, simply by virtue of its being a system.

But in a chimeric system we have further information. We know in advance that a chimera is a composite whose elements were themselves originally systems, or parts of other systems. Hence at some earlier time, these parts had their own identities, their own genomes, their own behaviors (phenotypes), and their own laws. Thus we presume a privileged set of parts, into which the behaviors (phenotypes) of the chimera as a whole can be *analyzed,* and out of which they can be *synthesized,* and likewise for the chimeric identity (or genome) that forces them.

Thus we come to focus attention on these parts and not the whole; to treat analysis and synthesis as mere inverse operations; and even more, as operations that preserve all identities (genomes) and the behaviors (phenotypes) they force. This is perhaps the essence of *reductionism*—that systems are to be related through the commonality of these parts, and only thus.

Conceptually, there is such a world, the reductionistic paradise, in which everything can be synthesized from fixed elements by following determinate rules (algorithms) expressed in a single coherent time frame from earlier to later. In this world, everything can be analyzed back into its constituent elements by following other algorithms. Anything in it is a mere sum of its elementary parts, and nothing new emerges. This is the world of machine or mechanism—what I called the world of *simple systems* in *Life Itself.* The belief that our real world of events and material reality is of this character is Church's Thesis. This kind of world is extremely feeble; in formal terms, one cannot do much mathematics in it; in material terms, one cannot do much biology, and hence not even much physics. In such a world, chimera cannot exist as a thing in itself; there are only more- or less-differentiated mixtures synthesized from, and analyzable into, elements of fixed and unchangeable identity, whose own identities are therefore fixed and unchangeable. Accordingly, the concept of *function* has no meaning. But again, such a world is very feeble.

Fortunately, there are rigorous alternatives that are far less feeble. I already touched on the concept of *models* in talking about states and parameters and system laws—i.e., about phenotypes and genotypes. A mathematical model of any system, of whatever character (even another mathematical system), is a formalism, whose inferential structure can be put into correspondence, or congruence, with what is happening in some other system or prototype. We can say either that the formalism *models* the prototype, or that the prototype *realizes* the formalism.

In some sense (see *Life Itself*), the totality of all behaviors of a system, its entire phenotype, is wrapped up in the totality of all models that can be made of it; this constitutes its epistemology. If, and only if, a system is simple, which means that all of its models are computable or simulable, then this set of all models becomes a reductionist paradise—otherwise, not. Stated another way, if a material system has a noncomputable model (i.e., it is *complex*), then its behaviors or phenotypes cannot all arise from algorithmic differentiation of a fixed finite set of elements, according to fixed identity-preserving rules or algorithms.

The following point of view serves to separate science from the reductionism with which it has (unfortunately) become identified. First, the general concept of a model allows us to relate or compare systems of utterly disparate character, not in terms of their parts but by the models they realize. This is the basis for system *analogy* (Rosen 1985a), which underlies such diverse activities as scale modeling in engineering, and the biomedical extrapolation of data (i.e., of phenotype) from one kind of organism to another (e.g., from rat to man) in biomedicine, and analog computation generally. So it potentially makes sense to compare and contrast, say, organic and technological systems, or organic and social systems, not because of the commonality of their material parts, but in terms of the models they realize. And, as it turns out, the concept of function constitutes the common currency for expressing this commonality.

Second, a complex system is one in which there must exist closed loops of entailment. Such loops cannot exist in a machine or simple system; this is indeed precisely why machines are so feeble (i.e., there is not "enough" entailment in a simple system to close such a loop). In mathematics, loops of this kind are manifested by impredicativities, or self-references—indeed, by the inability to internalize every referent. In science, where entailment means causality, closed causal loops (among many other things) allow us to talk rigorously about categories of final causation, divorced (if we so wish) from every shred of telos. This kind of finality, in turn, is what allows us to talk about function, and anticipation, in terms of what an effect entails in a complex system, rather than exclusively in terms of what entails the effect.

Finally, I return to the concept with which I began—that of chimera. If we are given a system, we can talk about its behaviors (its phenotypes) and the genotypes that force them in given environments. Each of these gives rise to a model of that given system. All this is utterly independent of the system's *ontology*—i.e., of where the system came from; it concerns only its *epistemology* as it exists. In a simple world, it turns out that the two coincide. In a complex world, on the other hand, they need not; we can know all about the one without knowing anything at all about the other.

What a model captures about the system it models has been expressed in many ways over the millennia; the old philosophers called it an *attribute* of that system, or an *essence* of it. The concept of system analogy allows us to compare diverse systems through their manifestation of a common attribute—e.g., a model that they both realize.

The expression of a particular system in terms of its attributes or models comprises a mode of *analysis* of that system (albeit not a reductionistic analysis). But in the world of complex systems, we cannot generally invert these modes of analysis to obtain corresponding modes of *synthesis*. Here, the realization problem becomes serious: Given an essence, or attribute (i.e., a model), the problem is to produce a material system that manifests precisely that essence or attribute (i.e., realizes that model). Because analysis and synthesis are not inverse operations in this world, we cannot automatically entail a realization (i.e., an existence) from an essence, or produce an ontology from a (posited) epistemology.

Particularly is this true if we want to realize several such models or essences simultaneously—i.e., if we want to fabricate a material system that realizes the given models. With respect to those models, any such simultaneous realization will be *chimerical*. On the other hand, solving such a realization problem is the very essence of technology—to make disparate essences coexist in a single system by producing an ontology for it.

Solving a realization problem, in the sense of producing an ontology for something that manifests a given epistemology (e.g., a given set of models of functions, a given phenotype) is very close to the creation of a life form. A *process* that solves such a realization problem will necessarily realize a basic functional model that is characteristic of life. To appreciate this, I will return to the notion of genome in this new context.

Genome in any kind of system, biological or not, is what conveys identity, or determines species, independent of all system behaviors. In biology, however, more is true: genome is what forces phenotype. That is, in biology, genome mandates behaviors (phenotypes) as well as identity. Moreover, phenotypes are where the *functions* are, where technology resides. A material system would not be called an organism, would not be recognized as "alive," unless these senses of genome coincided, unless what conveyed identity also forced its phenotypes, and hence determined their functions.

Now consider a system S that realizes a given model or essence, together with another system $F(S)$ that *fabricates* it. Together, they constitute a new (chimerical) system, which we may denote by $[S, F(S)]$. This new system has behaviors (phenotype) and genome (identity), as every system does. But it also satisfies the condition that the phenotypes are forced by the fabricator. This is characteristically biological. Thus if S embodies a technology, and $F(S)$ is a process that fabricates it, the composite of the two is in that sense "alive."

However, we can see an infinite regress forming. For we can ask What fabricates $F(S)$? Biology teaches us that we can avoid this regress by closing a causal loop; in biology, this constitutes replication. One cannot, however, close such loops in simple systems. Indeed, that is precisely what makes them simple (see *Life Itself*).

Finally, we return briefly to the hermit crab, the chimera, as the exemplar of technology. It exploits other genomes to realize functions that its own cannot. We do similar things: we can use, for example, a viral genome to make ourselves a geodesic dome, and thereby enlarge our own phenotypes. But behaviors that generate such chimeras, which augment functional capabilities and thereby create new life, are not themselves programmed in any sense. They embody at root the kinds of closed causal loops that programs are too feeble to encompass. Ultimately, the lesson of biology is that to seek survival through programming is, one way or another, a form of suicide.

CHAPTER 19

∾

Bionics Revisited

Many relations exist between biology and "the machine," or, more generally, between biology and technology. Certainly, the machine and machine technology impact on the concept of organism (Rosen 1987a, 1988a, 1989), but the interaction goes both ways: the organism is increasingly a source, or a resource, for technology. The interplay between the two was regarded as the domain of an ill-defined area once called bionics, not much of which is left now, and even that is vastly different from its initial conception.

St. Augustine said that God placed animals on the earth for four reasons: to feed man, to instruct man, to amuse man, and to punish man. We may say that man is putting machines on this earth for the same four reasons. But whereas God could survey His creation and see that it was good (and after He created man, He saw that it was *very* good), we are not yet in a position to make such a claim.

The merging of biology and technology, and, especially, the ultimate employment of organisms for technological means and as technological ends (as in, e.g., genetic engineering) presents a terrifying spectacle to many thoughtful observers. Consider, e.g., the words of Freeman Dyson:

> Whoever can read the DNA language can also learn to write it. Whoever learns to write the language will in time learn to design living creatures according to his whim. God's technology for creating species will then be in our hands. Instead of the crude nineteenth-century figure of Doctor Moreau with his scalpels and knives, we shall see his sophisticated twenty-first century counterpart, the young zoologist sitting at the computer console and composing the genetic instructions for a new

species of animal. Or for a new species of quasi-human being. . . . Can man play God and still stay sane? . . . The long-range threats to human sanity and human survival [come] from biology rather than from physics. (1979:169)

Are such concerns entirely fanciful? If they are not, to what extent are they offset by more dispassionate analyses, such as the one presented by Danielli:

The cost of an effective program . . . for transferring the nitrogen fixation gene set to crop plants . . . (and thus) of solving the food problem by biological means is expected to be two or three orders of magnitude less than its industrial equivalent. . . . Insofar as chemical industry can be replaced by biological industry, the upper limits to growth (imposed by industrial pollution, industrial wastes, and exhaustion of resources) can probably be raised by at least two orders of magnitude. (1974:4)

Such are the ultimate impacts of the interplay between biology and "the machine." We shall not, in the following, be directly concerned with such questions, but rather with another that illuminates them, and that can be phrased roughly as *Where (if anywhere) does machine end and organism begin?* Machine and organism are essentially different in kind, and, as a consequence, the concept of machine does not exhaust the dimensions of technology.

Some Historical Remarks

The idea of the organism as machine goes back to Descartes and the first decades of the seventeenth century, although as far back as Plato we find analogies between, for example, animal locomotion and the action of hinges or springs. Its allure has continually buttressed, and been buttressed by, the strategy of reductionism in biology. The current dogmatic faith in this metaphor accordingly rests on two pillars: the rise of molecular biology over the past half-century, and the concomitant growth of our own technologies during the same period. Some interesting ideas and developments have led to the present situation.

In the early 1930s, Rashevsky conceived the idea of the neural network as an explicit model for the brain (see chapter 7). This was an outgrowth of his earlier work on the physical basis of peripheral nerve excitation and nerve conduction. He was the first to show how quite simple networks could already manifest brainlike behaviors; he produced networks capable of discrimination, learning, memory, and the like.

Rashevsky's networks were described by systems of highly nonlinear differential equations and were accordingly difficult to analyze in any but the simplest cases. A decade later, McCulloch and Pitts (1943) produced a discrete Boolean version of these networks (see chapter 7). As purely algebraic objects, these were much easier to analyze. Moreover, it was immediately recognized that *any* binary Boolean switches could be used to construct such networks; hence we could begin to envision the fabrication of "artificial," or "synthetic," brains.

Such ideas seemed to join naturally with another stream, associated with, for example, Turing (1950) and von Neumann (1966). Turing had been concerned with numbers and computation, in light of developments (especially the Gödel Incompleteness Theorem) in the foundations of mathematics. The result was a class of "mathematical machines," the Turing machines, which embody the execution of algorithms.

Von Neumann had long been intrigued by those same foundational questions; moreover he, like Turing, participated decisively in the early development of digital computation. It was quickly realized that the "universal" machines of Turing were descriptions of general-purpose digital computers. It further became clear that the McCulloch-Pitts neural networks were themselves a special kind of Turing machine ("finite automata"); they became general Turing machines when attached to appropriate receptors and effectors. In this way, the distinction between brain and machine became blurred; many felt (and still feel) that it had disappeared entirely.

Around 1950, when these developments were coming to a boil, DNA was beginning to be identified with primary genetic material (and hence with genetic "information"). The structure of DNA, a linear copolymer, was irresistibly suggestive of an input tape to a Turing machine—i.e., as software to protein hardware. The cryptographic relation (i.e., coding) between a sequence of nucleotides in DNA and the corresponding sequence of amino acids in the resulting polypeptide chain was itself just a matter of symbol manipulation or word processing. Within a decade, the work of Jacob and Monod (1961) on differential gene expression led to the concept of the operon, and the cell itself as an operon network. This seemed to clear the way for an understanding of *differentiation.* In this picture, the operon is a formal unit that behaves like a Boolean switch; without knowing it, Jacob and Monod concocted operon networks identical with those of Rashevsky two decades earlier—networks that exhibited in that context brainlike behavior (see Rosen 1968).

Also around 1950, another circle of ideas developed, associated with the name of Norbert Wiener. He wrote *Cybernetics: Communication and Control in the Animal and in the Machine* (1948), and the wording of the subtitle was no accident. He had been interested in biology from childhood but had been discouraged from studying in this field, so he turned to philosophy and then mathematics. Then, as a result of his experience with automatic control, he felt he could return to biology from a different direction. He found natural relationships between his ideas of feedback regulation and the ideas just discussed.

Other men should be mentioned: Shannon and McCarthy (1956) and their Information Theory; von Neumann and Morgenstern and their Game Theory (1953); Pontryagin et al. (1961) and their Maximum Principle, and Bellman (1957) and his Dynamic Programming (these authors developed their powerful optimization techniques independently); and the Open Systems of von Bertalanffy et al. (1977). This was a very exciting time in the sciences.

In 1957, an apparently unrelated event occurred. The Soviet Union launched the first artificial satellite, bespeaking a technical level in rocketry and telemetry that the West had not matched. In the United States, the result was panic, and the consequence of that panic was the secretion of enormous, unlimited sums, almost haphazardly, to stimulate scientific research (and, in particular, to close the "missile gap"). This lasted for a decade (roughly 1958–1968). A good bit of the money flowed into what were then called the inter-disciplines, which enjoyed a brief efflorescence. One of these was called bionics.

As initially conceived, bionics involved the interplay between biology and human technology in their broadest sense. Its twin goals, insofar as they were articulated, were (1) the employment of biological modes of behavior and organization to solve technological problems, to design new and better ways to engineer, and (2) to use technology (e.g., holography) to illuminate biological processes themselves. Thus, initially, biology and technology were viewed as co-equal partners in this enterprise.

In those days, bionics was viewed as an urgent imperative. There was (and still is) a prevailing feeling that our technology was too simple-minded for the problems it was increasingly being called on to solve, that this gap between our problems and our capacity to resolve them was increasing rapidly, and that we needed therefore somehow to tap into the billions of years of experience in coping with "complexity" that evolution provides. It was the goal of bionics to make this encyclopedia, written by natural selection, an ongoing resource for humanity—a resource that would manifest itself in everything from prosthetic devices (e.g., an artificial hand that could be used to play the piano) to the "intelligence amplifiers" of Ross Ashby.

I well remember the dozens of meetings and symposia concerned with bionics (often referred to as self-organization) during that decade. And that was just in the United States. The same thing was occurring in England, in Europe east and west, and in the Soviet Union. It seemed in those days that a critical mass had been achieved, and that bionics had already become a permanent fixture of the scientific landscape. Tangible evidence of this was provided by the representatives of granting agencies who prowled the floors of such symposia in search of investigators to give money to.

One of the main foci of bionics then was the interplay between brain and computer, which seemed to provide a natural playground for bionic ideas. Developments such as Frank Rosenblatt's (1962) perceptron, and its variants, promised to throw a new light on both software (programs) and hardware (the wiring of a neural net). It seemed then as if machines that think were just around the corner. Out of this ferment crystallized what is now called AI (for artificial intelligence), the only remnant of bionics that continues to exist in coherent form.

In the early days of AI, there was as much emphasis on real brains, on neurophysiology, as on machines or programming. But, especially as resources rapidly began to dwindle in the late 1960s, the field came more and more under the domination of those who, in their arrogance, felt they could do it better than Nature, and who accordingly discarded the biological side as irrelevant and unnecessary. Who cares, they said, how people play chess, or learn languages, or prove theorems? As Marvin Minsky, one of the more articulate apostles of this viewpoint, is supposed to have put it, one does not learn to build an airplane by watching a bird flap its wings.

In any event, AI is all that really remains of that initiative of bionics three decades ago, and even that exists only in vastly transmuted form. The fundamental shift in concerns that characterized the late 1960s, the enormous contraction of the base of scientific support that accompanied it (and, in particular, the dismantling of the military research agencies that had mainly funded bionics), and the disbanding of the numerous research "teams" that had been formed to pursue interdisciplinary programs simply killed it.

The Airplane and the Bird

The concerns that initially animated the development of bionics, and the problems with which it tried to deal, have of course not gone away. In many ways, they have become sharper and more urgent, while at the same time we increasingly lack the coherent response to them that bionics tried to provide.

What do these problems look like today, after the passage of more than thirty years? In particular, what is the current perspective on the interplay between biology and the machine, and more generally, between biology and technology? What, indeed, is the relation between the airplane and the bird, and what are the lessons to be learned from it?

In reductionistic world-view as it existed in the mid-1950s, physics encompassed all of material reality—everything from fundamental particles to the universe. Every material system was subject to its analytical modes and to the universal laws entailed by them. Complete explanation, complete understanding of any material system consisted of this and of this alone.

In particular, organisms and machines comprise definite (if ill-defined) classes of material systems. The machine metaphor says much more than this: it says that the class of organisms is contained in the

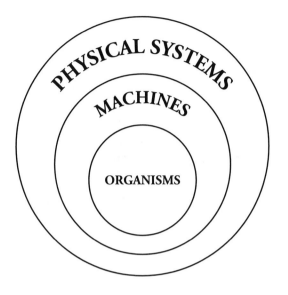

FIGURE 19.1

class of machines. And physics says that the class of machines is contained in the class of material systems. We thus have a taxonomy—the nest diagrammed in figure 19.1.

One further thing: von Neumann, and others, had drawn attention to a notion of complexity, which they felt was important for such a taxonomy or classification of material systems. Roughly speaking, this complexity was measured by a number or numbers that could be effectively associated with any such system, in such a way that the system would be called *simple* if these numbers were small, and *complex* otherwise. Von Neumann in particular argued that there was a *finite threshold of complexity;* below this threshold, we find the machines of ordinary experience, which could only deteriorate; above the threshold, we find machines that could learn, grow, reproduce, evolve (i.e., could do the things that only flesh is now heir to). Crossing this threshold, then, was tantamount to creating life, and complexity in this sense became an explanatory principle for the characteristics of life.

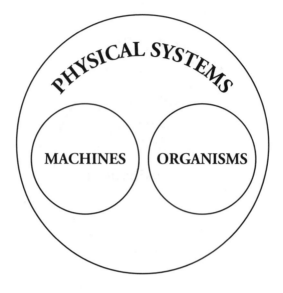

FIGURE 19.2

Amending figure 19.1 to take account of this notion of complexity results in figure 19.2. In this diagram, all the organisms are to one side (the *complex* side) of the complexity threshold.

This picture entails many things. I shall state a few of them. First, it entails that (to use a philosophical but entirely appropriate language) *ontology and epistemology coincide.* That is, it entails that an understanding of how something works also tells you how to build it, and conversely. As von Neumann put it, *construction and computation are the same thing.*

Second, a related assertion: *Function is always localized in structure.* That is, functional activities in a machine can be spatially segregated from one another by artificial means, without loss of function. Indeed, the efficacy of reductionistic analysis absolutely requires this property— a property that I shall call *fractionability.* Fractionability *must* hold ubiquitously, whether our system is simple or complex in the sense just described; otherwise ontology and epistemology need not coincide, and hence all the inclusions that characterize figures 19.1 and 19.2 cease to hold.

According to these suppositions, an organism is a gadget—a piece of technology. And bionics was concerned with a technology transfer between organisms and other machines. More specifically, it was intended to go from organism epistemology to machine ontology, *across the presumed complexity threshold* (from complex to simple). Conversely, the effect of technology on biology was envisaged as going in the opposite direction, from simple to complex—from machine ontology to biological epistemology. And crossing that threshold (from simple to complex) amounts to the creation of life, the ontology of organism.

The bionics of thirty years ago tacitly accepted everything in the diagrams of figures 19.1 and 19.2—the universality of physics, the machine metaphor, the threshold of complexity, all of it. Accordingly, it accepted everything entailed by these diagrams.

Now, what of the petulant assertion with which artificial intelligence dismissed its bionic heritage: *If you want to build an airplane, you don't watch a bird flapping its wings.* Is this assertion intended to be the statement of some kind of principle? Or does it rather bespeak only a description of our present impoverished abilities to hold the mirror to Nature, as it were?

The assertion is, in fact, about *fractionability.* It is, first of all, an assertion regarding separation, or segregation, of the flight from the bird. This kind of fractionation is what I have called (1978b) *alternate realization*—in this case, of a function (flight) by distinct and otherwise unrelated material structures (bird and airplane).

Further, what makes the bird and the airplane so different in this case is again a matter of fractionability. The bird wing, for example, is an *unfractionable* combination of engine and airfoil. We cannot *physically* segregate these two functions into separate spatial structures. In the airplane, on the other hand, engine and airfoil *are* fractionable one from the other. They are fractionable because *that is how the airplane is built*—it is a consequence of its own ontology.

This last observation is important. We generally construct things sequentially, by accretion, one structure at a time, one function at a time. That is how we build a machine; that is, in fact, *the only way we know how* to construct anything. Accordingly, we cannot build a bird; its ontology is different from that, and not equivalent to it. We *don't* build a bird wing, ultimately, because we *can't* build a bird wing; its nonfractionability removes it instantly from the province of things we can build.

In considerations such as these, we find the true lessons of biology regarding machines and technology—that is, for the embodiment of functions in structure. Biology is telling us some terrible, disconcerting things: it is telling us that synthesis equals analysis, that ontology equals epistemology, that computation equals construction, that syntax equals semantics, and a host of other radical things of the same character. Biology is telling us that there is something inordinately *special* about a world in which all these inequalities are replaced by equalities: such a world is full of degeneracies, of *nongenericities*.

Minsky's dictum regarding birds and airplanes, however, is a cheery assurance that the nongeneric, simple world of the machine is nevertheless *good enough*. A neural net may not *be* a brain, nor an airplane a bird, but they are nevertheless *just as good*. And moreover, we can build them without having to learn anything new, or doing anything different; that makes them *better*.

The true lesson of biology, however, is that the impoverished world of machine and mechanism is not just as good, any more than mere symbol manipulation is just as good as Number Theory, or mathematics, or language itself. I shall turn briefly to these matters.

On Complexity

A system is *simple* if all its models are simulable. A system that is not simple, and that accordingly must have a nonsimulable model, is *complex*.

This usage of the term *complex* is completely different from that employed heretofore. This is unfortunate, but there were no other terms that could be used. In von Neumann's terminology, *every* system is simple in my sense; what he calls complex I would merely call *complicated*.

The world of systems that are simple in my sense is the world of *mechanisms*. It is the world with which contemporary physics concerns itself. It is also the world in which all the degeneracies or nongenericities mentioned above actually hold. It is a very nice, tidy, orderly world, this world of mechanisms. And it is a closed world, in the sense that once we are inside it, we cannot get out—at least not by internal, purely syntactic operations alone. In that way, it is very like the mathematical universe of finite sets (indeed, there are many profound parallels between cardinality of sets and complexity).

Using this terminology, I suggest a taxonomy for natural systems that is profoundly different from that of figure 19.1 or figure 19.2. The nature of science itself (and the character of technologies based on sciences) depends heavily on whether the world is like figure 19.2 or like this new taxonomy.

In this new taxonomy, there is a partition between mechanisms and nonmechanisms. Let us compare its complexity threshold with that of figure 19.2. In figure 19.2, the threshold is porous; it can be crossed from *either direction,* by simply repeating a single rote (syntactic) operation sufficiently often—an operation that amounts to "add one" (which will ultimately take us from simple to complex) or "subtract one" (which will ultimately take us from complex to simple).

In the new taxonomy, on the other hand, the barrier between simple and complex is not porous; it cannot be crossed at all in the direction from simple to complex; even the opposite direction is difficult. There are certainly no purely syntactic operations that will take us across the barrier at all. That is, no given finite number of repetitions of a single rote operation will take us across the barrier in either direction; it can produce neither simplicity from complexity, nor the reverse.

As I have pointed out (1988a), the Gödel Incompleteness Theorem (Gödel 1931) may be looked upon as an assertion that Number Theory comprises a (purely formal) system that is complex in our sense. Number Theory cannot be formalized (i.e., reduced to software to some Turing machine and thus expressed as a purely syntactic game of symbol manipulation), without losing most of its truths.

Let us look at this situation in an ontological perspective. In a formalization, computation and construction are the same; epistemology equals ontology. But construction (symbol manipulation) in this context involves the application of rote syntactic operations (production rules) *in a given sequence* (i.e., in a single fixed time frame). This is, in fact, what an algorithm is. Whatever can be constructed in this way is sometimes called *predicative.* The whole point of formalization was to try to say that *everything is predicative.*

Impredicativity, on the other hand, was identified as the culprit in the paradoxes springing up in Set Theory. Something was impredicative if it could *not* be constructed by applying rules taken from a finite set of syntactic productions, applied in a definite order to symbols drawn from a fixed set. In particular, something was impredicative if it could be defined only in terms of a totality to which it itself had to belong. This, it was held, creates a *circularity:* what is to be defined could be defined only in terms of a totality, which itself could not be defined until that element was specified.

Formalizations are simple systems (in my sense) and, in particular, cannot manifest impredicativities or self-references or "vicious circles." This is precisely why such a simple world seemed to provide a mathematical Eden, inherently free from paradox and inconsistency. Alas, as Gödel showed, it was also free of most of mathematics. We cannot dispense with impredicativity without simultaneously losing most of what we want to preserve.

Looking at this notion of impredicativity with ontological eyes, we see that there is no algorithm for building something that is impredicative. Thus in a world in which construction equals computation, an impredicative thing cannot be made to exist; it can have no ontology.

Note that *impredicativity and what I called nonfractionability are closely related.* Indeed, the latter is essentially an ontological version of the former. In the example used earlier, for instance, we cannot fractionate airfoil from bird wing—that is, we cannot define the airfoil except in terms of the wing, nor can we define the wing without the airfoil— a typical instance of impredicativity.

Even in physics, the science of the simple, impredicativities continually arise and invariably give trouble. They arise in the context of self-referential properties, which immediately set up the characteristic "vicious circle." This, indeed, is the ultimate source of the infinities of "self-energies" that have haunted, for example, electrodynamics and quantum field theory. From our present viewpoint, infinities are the necessary consequence of trying to cope with an impredicativity by purely simple, formal, syntactic means.

In a nutshell, we find that impredicativities (i.e., complexity) and pure syntax are incompatible. More specifically, complexity and an ontology based on a single syntactic time frame (the ordering of purely syntactic operations into individual steps) are incompatible. In the present context, in which we have identified impredicativities with nonfractionabilities, we cannot build nonfractionable systems by purely syntactic means either. We must accordingly either invoke semantic elements transcendental to syntactics (e.g., taking of limits) or (what may perhaps be equivalent) utilizing two or more *incommensurable time frames.*

Some Perspectives for Tools and Technology

In the relatively innocent world of forty years ago, the world in which bionics was first conceived, none of the matters we have touched on here had even been dreamed of. Everything was simple, and fractionable, and simulable; technology was machine, and organism was machine, and bionics was to be the systematic exploitation of these facts to mutual advantage.

The considerations I have developed here suggest quite a different picture. Machine is simple, organism is complex. Thus the machine metaphor is false: organism and machine are different in kind.

If organism is not machine, then technology need not be machine either. Perhaps, indeed, it dare not be machine, in the context of social and biological technologies—i.e., technologies inherently associated with complexity. This was a possibility that could not have occurred to those concerned with bionics in the early days.

All that is left of that innocent identification of organism and machine is the fact that complex systems can have simple models (Rosen 1989). Thus the bird wing *can* be modeled as airfoil. But that model is impoverished, and one manifestation of that fact is that the airfoil model of bird wing can also be realized by nonbiological systems—i.e., by simple systems such as airplane. But if we go deep enough into the category of models of bird, we must find some models that cannot be realized by *any* simple system (e.g., by airplane)—just as Number Theory has theorems that cannot be reached by any formalization.

The True Scope of Bionics

I suggest that the ultimate lesson to be learned from comparing the bird wing to the airfoil lies in the realm of complexity: The bird is an organism and is different in kind from the machine. We can learn about the machine as a simple manifestation of the complex organism, but the reverse does not work. In causal terms, the machine must always be supplied with aspects of final causation; as Aristotle pointed out two thousand years ago, a machine cannot be understood unless we can answer *why* questions about final causation. These questions supply the complexity inherently missing from the machine, except that the complexity refers now to ourselves as understanders and not to the system being understood. Only thus can bionics play the intermediate role between machine and organism that was formerly visualized for it.

ɔ

On the Philosophy of Craft

Medicine has been called an art and it has been called a science. Indeed, it must possess elements of both. But primarily, it is a craft in its practice, and a technology in its aspirations. It is applied science—primarily applied biology. In some of its aspects, it is even rather more applied technology than applied science.

Science has always had philosophy associated with it; indeed, for a long time, science was called natural philosophy. The ancient Greeks were keenly interested in the way the world was put together and how it worked, and they had laid out the major alternatives in this connection (e.g., atomicity or infinite divisibility; evolution or special creation) two millennia ago. But, perhaps because the Greeks affected to despise craft and considered technology to be the province of slaves, there has never been much of a philosophy of either craft or technology in general.

Indeed, one would be surprised to find a work entitled "The Philosophy of the Airplane" or "The Philosophy of the Automobile," and even more so to find one entitled "The Philosophy of Automobile Repair and Maintenance." The former involve technologies; the latter is a craft. Although a medical doctor would surely resent being analogized to a repairman, there are powerful grounds for doing just this. Indeed, contemporary biological science is currently locked in the grip of a Cartesian tradition that asserts that organism *is* machine—nowadays perhaps qualified to read "molecular machine," but machine nevertheless.

If one accepts this machine metaphor, which is one of the primary underpinnings of contemporary reductionism in biology, then an organism becomes a piece of engineering (albeit without an engineer), and a physician is one responsible for the maintenance and proper functioning of this piece of engineering. Indeed, in this light, medicine itself becomes a species of engineering, of a kind nowadays called control engineering.

If a machine malfunctions (i.e., exhibits pathology) or otherwise deviates from nominal behavior, one must seek the causes for the aberration (troubleshooting, or diagnostics) and then make the necessary repairs or adjustments (therapeutics or prosthesis). Even in the absence of such pathology, one must generally maintain the machine in such a way that nominal behavior will be preserved (preventive maintenance, or hygiene).

The only difference between medicine and other forms of control engineering is that, since we neither built the machine we call an organism (i.e., we are not responsible for its ontology) nor know much about how it actually works (i.e., its epistemology), the physician is much more in the dark than is his technological counterpart. Instead of nominal performance criteria, he must make do with a notion of health, which has proved difficult or impossible to quantify completely. Instead of explicit troubleshooting protocols based on design, he must rely on a restricted set of diagnostic procedures that rest ultimately on experience. The same is true for therapeutic procedures, most of which create further problems (side effects) requiring further therapies (or iatrogenics). About the only advantage the physician has is that his "machine" can often heal itself, and it can report to him where it hurts.

Because of these intrinsic uncertainties, philosophy reenters the picture, even if one accepts the machine metaphor (itself, of course, a philosophy of organism). We must ask questions such as, What is health? or, equivalently, What is pathology? What is disease? What are symptoms or syndromes? What is a therapy? How can we minimize or eliminate therapeutic side effects?

We can frame such questions whether or not we accept the machine metaphor. But the kind of answers we come up with, or even can hope to find, differ radically if the machine metaphor itself is wrong. I have come to believe the latter, for many reasons. I shall try to describe some of them, in a roundabout way, in what follows, and to draw some conclusions pertinent to our subject.

Magic Bullets and Side Effects

I shall illustrate some of the basic ideas involved by considering the notion of *side effect,* by which I mean an unforeseen consequence of a therapeutic intervention undertaken for another purpose. A therapy without side effects used to be referred to as a *magic bullet;* it accomplishes
its nominal purpose without any other effect on the system. In dealing
with mechanical or electrical devices, all effective therapies are magic
bullets. If the machine metaphor is correct, we would expect that, ide-
ally, medical magic bullets would always exist as well, at least in principle. But our experience is quite otherwise.

Why is this? It could be, of course, that we are simply too ignorant
of the details of organism, and the effects of interventions, to find the
promised magic bullets. Or on the other hand, it could be that the ubiquitous side effects we find are themselves telling us something new
about organisms—something basically incompatible with the machine
metaphor itself. I shall briefly sketch an argument for the latter possibility.

Suppose we are sitting in a big room, with a lot of air in it. Since
there is a lot of air, the temperature of the room is not too sensitive to
external influences. So our short-term experience will be that the temperature of the room does not vary, and we extrapolate that this temperature will remain unvarying. And suppose we like the temperature as
it is.

Eventually, of course, the temperature will begin to change, because
the room is open to ambient influences we do not see directly. That is,
the actual room temperature is departing from the expected and desired
temperature. Of course, we think: noise. The diagram of this situation,
shown in figure 20.1, is familiar from any textbook on control theory.
The noise, being outside, is *unpredictable:* we have no model for it. But
we want to eliminate its effects; that is, we want to restore the constancy
of the room temperature and maintain it at the value we like. So what
therapy shall we institute?

Well, if we are very clever, we will build a thermostat, a feedback
regulator that will sense the room temperature, compare it to the desired
temperature, and institute a corresponding control. That is, we will em-
bed our original system in a larger one, of the form in figure 20.2. We
have thereby "cured" our room; we have created a kind of dynamic insu-

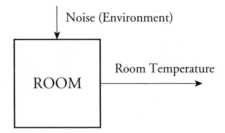

FIGURE 20.1

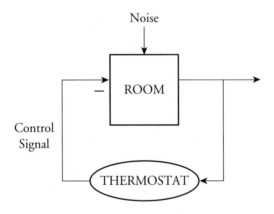

FIGURE 20.2

lator that *closes* our room from the effects of ambient temperature. And formally, we have created, in the changing behavior of the control loop itself, a representation of the *noise* that was initially lacking.

But we are not done yet. The thermostat itself is new material structure, which we have had to bring into the system to control the effects of unpredictable temperature fluctuations. The thermostat closes the room off to ambient temperature, but it itself is now open to other interactions—i.e., to new sources of noise. For instance, parts of it may corrode because of humidity and oxygen of the air in our room. These new sources of noise may interfere with its action *as a thermostat,* and hence ultimately with the original temperature control itself.

Thus we must pay for closing off the original system to ambient temperature by opening the system to new sources of noise. This is a typical kind of side effect, arising already here in a purely artifactual situation. Diagrammatically, instead of the closed loop of figure 20.2, we must instead have the open situation of figure 20.3, where we may indeed end up with more noise than we had originally.

What to do? We obviously need more therapy to control the side effects we generated by instituting the original therapy. We could simply iterate what we did in going from figure 20.1 to figure 20.2—i.e., put in more control loops to eradicate the new sources of noise. We would then get something like figure 20.4.

It should be clear what is happening here. At each stage of the game, we institute controls (therapies) to handle the side effects of the previous stage. But by doing so, we only open the system up further.

Even in this simple example, we see an incipient and deadly infinite regress yawning before us. Indeed, this little example illustrates a situation that can be formulated in very general terms. The real question arising here is whether, and if so, when, this potential infinite regress can be broken off. And as we shall now briefly indicate, that question constitutes a probe of the machine metaphor itself.

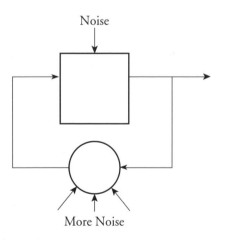

FIGURE 20.3

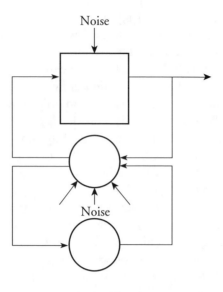

FIGURE 20.4

Coping with Infinite Control Regresses

Let us look again at the first stage in our incipient infinite regress, at the little homeostat diagrammed in figure 20.2. The controller is *designed* to provide, in the control signal it emits, an exact image or model of the noise (i.e., of the *environment*) that is affecting the behavior of our system. This control signal is fed back negatively into the system itself, so that

$$\text{Control signal} + \text{Noise} = 0. \tag{20.1}$$

When this condition is satisfied, the *set point* of the controller becomes a model of the *system,* or

$$\text{Temperature} = \text{Set point} = \text{Constant.} \tag{20.2}$$

But the controller itself is subject to noise. The result is that equation 20.1, and hence equation 20.2, will fail to be satisfied. In other words, the control signal of figure 20.2 is no longer a model of the original noise we wanted to control.

Hence the second control loop of figure 20.4. The combined control signals of the two loops in this figure provide a better model of the original noise perturbing our system than does the first loop alone. But it is still not good enough, so we must iterate again, and yet again, indefinitely.

However, it is conceivable that such a potential infinite regress actually breaks off. It will do so if, and only if, we can arrange matters so that the noise arising at the Nth step of this sequence produces consequences that are subject to controls instituted at earlier stages. In such a case, the sequence breaks off at the Nth stage. Indeed, this finite sequence of control loops, of length N, constitutes our magic bullet. More explicitly, the control signals of the N loops together constitute a complete model of the original noise (i.e., of the original system's environment), and the set point of the first loop is, in effect, a complete model of our original system.

In this example, the system we wished to control could be characterized by a single state variable (temperature). Moreover, there was no autonomous dynamics; there was only a response to noise. It is not hard to extend our analysis to much more general situations, in which the system we wish to control must be characterized by any finite number of such state variables (or explicit functions of them—i.e., observables), each one under the influence of the others (the autonomous dynamics) and of external noise (the forced dynamics). In such a case, each state variable, and each parameter of the autonomous dynamics, can be separately treated as in our little example. The analysis, of course, grows increasingly complicated, but, in effect, we now have a much larger family of cascading control loops, each of which creates the potentiality for infinite regress.

A situation in which every such cascade breaks off after a finite number of steps turns out to be equivalent to what I call *simplicity*. I define a system to be *simple* if all of its models are computable, or simulable. In such a case, and only in such a case, the totality (or *category*) of all models contains a unique maximal element, a biggest model, from which all other models can be extracted by purely formal means.

Another name for such simple systems is *mechanisms*. Indeed, in such a case, there must also be a spectrum of minimal models ("atoms") out of which the biggest model can be canonically constructed. The class of simple systems is the reductionistic paradise; the basic assertion of reductionism is that *every material system is simple.* This is just a material version of Church's Thesis.

Now, a sequence of cascading control loops of this kind constitutes a model of the system being controlled, and its environment. Accordingly, (1) if these controllers are themselves simple, and (2) if every such cascade breaks off after a finite number of steps, then the system itself, and its environment, *must both be simple.* Conversely, if a system (or its environment) is not simple, then there must be at least one cascade of simple controls that does not break off. Or, if there is such a cascade that does not break off, then the situation we wish to control cannot be simple.

A system that is not simple in this sense (i.e., is not a mechanism) I call *complex.* (My usage of the term *complex* differs essentially from the way other authors define it. In particular, simplicity and complexity are not things that can be measured by numbers; they are not merely additional numerical parameters that give *simplicity* when small and *complexity* when large.) The machine metaphor says that *all organisms are simple* in this sense. The truth or falsity of the metaphor will have profound implications for medicine as craft, and indeed for craft and technology in general.

An Aside on Complex Systems

I am suggesting that what I call complexity is in fact the normal or generic situation; accordingly, a gratuitous hypothesis of simplicity (i.e., computability, or simulability, of all models) is an excessively strong one, locking us into a very special universe. I shall briefly consider a cognate situation that has arisen in mathematics and then sketch its relation to the main subject under consideration.

Mathematics itself, over most of the past century, has been in the throes of a foundations crisis, precipitated in part by what happened to Euclidean geometry and in part by the discovery of paradoxes in Set Theory. The specific nature of these need not concern us, beyond noting their joint preoccupation with consistency of formal systems of inference, and that the issue is still wide open.

One attempted resolution of this crisis was proposed by David Hilbert in the early decades of this century, and was called formalism. In a nutshell, Hilbert proposed to regard all of mathematics as merely a game played with a finite family of meaningless symbols, manipulated according to a family of fixed rules. That is, a mathematical system becomes a game of pattern generation—of symbol manipulation or word processing—and, as such, it could be carried out entirely by a machine. Indeed, machines of this kind were subsequently characterized by mathematician Alan Turing. These Turing machines were then further identified with digital computers. The terms *computable* and *simulable,* which I used previously, refer precisely to what such machines can do; roughly, something is simulable, or computable, if it can be expressed as software (e.g., as program) to the hardware of such a machine.

Hilbert's program of formalism asserted that any mathematical system could be thus expressed, without any loss of content. It thus constitutes a formal version of the reductionistic hypothesis, that every (in this case mathematical) system is simple.

The Hilbert program was wrecked in 1931 with the publication of the celebrated Gödel Incompleteness Theorem. In effect, Gödel showed that the Theory of Numbers, perhaps the innermost heart of all of mathematics, could not be formalized in this sense . . . unless it was already inconsistent. Gödel showed explicitly that any attempt to formalize the Theory of Numbers necessarily lost most of its truths. That is, he showed that *the Theory of Numbers was complex* in our sense.

Now, there is a deep parallel between the formalist view of mathematics (i.e., of meaningless symbols pushed around by formal rules) and the reductionistic view of the material world as consisting of configurations of structureless (meaningless) particles pushed around by impinging forces. And just as Hilbert sought to replace all modes of mathematical inference by symbol manipulation, so the reductionistic view seeks to capture all causality in laws that tacitly amount to the computability of all models.

Computability in mathematics is an excessively strong stipulation that characterizes a vanishingly small class of unusual (and very weak) systems. And the same is true in the material world as well: there is much more to causality (the material analog of mathematical implication or inference) than what goes on in simple systems.

Mathematicians do not stop doing Number Theory because it is not formalizable. They simply view formalizability as an irrelevant impediment. Likewise, we need not stop doing science when confronted with complexity. But it does change the way we look at science, and hence the way we look at crafts and technologies associated with it.

Some Preliminary Conclusions

I have attempted to introduce, and to motivate, a concept of complexity. A system is called complex if it has a nonsimulable model. The science of such complex systems is very different from the science we have become used to over the past three centuries. Above all, complex systems cannot be completely characterized in terms of any reductionistic scheme based on simple systems. Since the science is different, so too are technologies based on it, as well as any craft pertaining to the systems with which that science deals.

In particular, we must adopt a vastly different view of medicine if we assume that the organisms it treats are complex systems rather than simple ones (i.e., machines or mechanisms). Complex systems are much more generic than simple ones, just as, say, transcendental numbers are more generic than rational ones. In particular, computability of system models is, in itself, an inordinately special thing to require.

Contemporary physics is replete with such nongenericities. Conservation laws, closure, symmetry conditions, the exactness of similar differential forms, and a host of other familiar things of similar character guarantee nongenericity, and at root are consequences of a prior presumption of simplicity.

In general, nongenericities involve degeneracy—the forcing of two initially distinct and independent things into coincidence, or the imposition of relations on things initially unrelated. So it is here: simplicity creates many such degeneracies.

For instance, one of the corollaries of simplicity can be stated as follows: The ontology of a simple system is entirely subsumed under its epistemology. In contemporary biology, for example, it is supposed that the same set of analytic units that supposedly accounts for the operation or functioning of an organism also somehow accounts, simultaneously, for its fabrication or origin. This is a direct legacy of the machine metaphor: the same set of parts that allows us to understand the operation of a watch, say, also allows us to (re)assemble it, and even more, to diagnose, and repair, malfunctions in its operation.

In complex systems, this is no longer the case; in complex systems, ontology and epistemology are generally very different. Our contemporary science, which has concentrated almost exclusively on epistemological issues, accordingly gives very little purchase on ontological aspects when the two are different. That is why the origin-of-life problem (or for that matter, the origin-of-anything problem) is so hard.

Since medicine in its diagnostic mode involves essentially epistemology, whereas in its therapeutic mode it involves ontology, the separation of the two in complex systems becomes a central matter.

Indeed, the entire concept of craft changes completely when dealing with complex systems. For we cannot generically approach them exclusively by simple means. There is a sense in which complex systems are infinitely open; just as with any infinite thing, we cannot exhaust their interactive capacities by attempting to control their parameters one at a time. In particular, the simple control cascades previously mentioned will generally not break off; hence, magic bullets of this character will generally not exist.

On the other hand, we are now free to envision control strategies involving controllers that are themselves complex. On the face of it, there is nothing to prevent such controllers from constituting the desired magic bullets. In fact, for complex systems in general, complex controls offer the only reasonable hope.

It is too early to tell how such ideas will develop in the future. My purpose here has been to introduce some of the flavor of the concept of complexity, how it pertains to basic biological issues, and how it may force a complete reevaluation, not only of our science, but of our concepts of art and of craft as well. Indeed, it may turn out, as it has before, that the pursuit of craft may provide the best kind of probe to guide our science itself.

CHAPTER 21

༄

Cooperation and Chimera

In mythology, Chimera was a monster built up from parts of lions, goats, and serpents. According to Homer, it breathed fire. It may have represented a volcano in Lydia, where serpents dwelt at the base, goats along its slopes, and lions at the crest. Accordingly, *chimera* has come to mean any impossible, absurd, or fanciful thing—an illusion.

In biology, aside from being the name of a strange-looking fish, *chimera* refers to any individual composed of diverse genetic parts—any organism containing cell populations arising from different zygotes (McLaren 1976). Indeed, the fusion of two gametes to form a zygote may be looked upon as a chimerization—a new individual produced from parts of others. This is a kind of inverse process to *differentiation,* in which a single initial individual spawns many diverse ones, or in which one part of a single individual becomes different from other parts. Chimera formation has been looked upon as aberrant: a curiosity or pathology when it occurs spontaneously, and mainly involving directed human intervention when it does not. (Grafting, in all its manifestations, provides an immediate example.) Yet chimeras are everywhere around us: ecosystems, social systems, and man-machine interactions; even chemical reactions can be thus regarded.

Here I shall look at the evolutionary correlates of chimera formation, and particularly at chimera in the sense of an adaptive response, based on modes of cooperative behavior in a diverse population of otherwise independent individuals competing with each other as such. If we take chimera seriously, in the sense of being a new *individual* with an identity (genome) and behaviors (phenotype) of its own, we can raise some deep epistemological and system-theoretic questions. These range from the efficacy of reduction of a chimera to constituent parts, all the way to sociobiology (i.e., what *is* phenotype anyway?) and beyond.

To illustrate these questions, let us consider again the hermit crab discussed in chapter 18. It is a true natural chimera, part arthropod (the crab itself), part mollusk (its adopted shell), and part echinoderm (the anemones that grow on that shell)—the *pluribus*. The parts together form a single functioning whole, with its own behaviors and its own identity—the *unum*. But what is its genotype? What is its phenotype? How are the two related, both to each other and to the identities and behaviors of its constituents? How did it evolve? What, in short, are the relations between the initial pluribus and the resulting unum?

Some Generalities About Evolution

Nowadays, the word *evolution* is used very carelessly. We talk about the time evolution of an arbitrary dynamical process, as if evolution were synonymous with change of state and hence with arbitrary dynamical behavior. But this usage is simply an equivocation that permits anyone studying any kind of dynamics to call himself an evolutionist, and to say (vacuously) that everything evolves.

From the very beginning, however, the word *evolution* was inextricably linked to certain modes of *adaptation*. What is it that qualifies a particular dynamical behavior, a change of state, as adaptive? The term *adaptation* requires the prior stipulation of that individual whose survival, in some sense, is favored or enhanced by it (Rosen 1975). Unless we so stipulate, a change of state, or behavior, can be adaptive for a subsystem but maladaptive for a larger system that contains it, and vice versa. Thus, for instance, a neoplastic cell may well be more adaptive than its normal counterparts, but neoplasia is highly maladaptive for the organism to which the cell belongs. Conversely, restriction of adaptive capability, which is often bad for a given cell, favors the survival of the whole organism. Likewise, what favors a species may kill its ecosystem. What favors a firm might kill the economy. Indeed, it is seldom true that, as a president of General Motors once asserted, "What's good for GM is good for the country." Garrett Hardin (1977) has called such situations "tragedies of the commons."

The contrapositive of this attitude was stated by Dr. Pangloss in Voltaire's *Candide* (lampooning Leibnitz): "Private misfortunes make up the General Good. Therefore, the more private misfortunes there are, the greater the General Good."

The "invisible hand" that rules over this kind of cacophony, in the context of biological evolution, was called by Darwin *natural selection*. Its seat was placed not in the adapting individuals themselves, but rather in their environments (which of course included all other individuals). In Darwin's view, this "hand" was based on, and driven by, competition—the struggle for existence. This view necessarily assigns to *environment* a number of distinct roles. First, the environment was the repository of what its resident individuals were struggling *for*. At the same time, it was also the repository of what each individual was struggling *against*. And finally, it made the decision as to who was winning and who was losing—a judgment as to *fitness*.

Our basic question is, In what sense is a chimera, a cooperative thing, an adaptive response in an evolutionary context? How could such a thing arise, let alone prosper, in a situation presumably driven by competition? Our point of departure will be the statement that to qualify as an adaptation, a behavior must enhance the survival of the individual manifesting it. There are a number of vague terms in this characterization—*individual, survival,* even *behavior*—and we must make these heuristic terms more precise. To do so requires digressions into epistemology and ontology.

Some System Theory

The *mechanism* that most biologists espouse provides definite answers to some of our questions. It stipulates what an *individual* is and what *behavior* is, but what *survival* means does not have any straightforward mechanistic counterpart. Yet, as we have seen, *adaptive* behavior is linked to *survival*, and Darwinian evolution depends on adaptation. Biologists, in general, feel that they dare not stray from mechanistic strictures without falling into vitalism. Hence the conundrums that have always plagued what passes for evolutionary theory (see *Life Itself*).

I shall first discuss the mechanistic representations of individuals and their behaviors. A behavior, in general, is represented as a change of *state.* What changes a state is the assignment of a tangent vector to the state, and interpreting it as the temporal derivative of state. Now, a tangent vector, like any vector, has both magnitude and direction. The *direction* is determined by, or represents, the *forces* that are posited to be responsible for the behavior; the magnitude is determined by what I shall call *inertial properties,* or inertial parameters, that modulate between the posited forces and the system behaviors elicited by them.

In terms of Aristotelian causality, when we treat the resultant behavior as an *effect* and ask Why? about it, we get three answers:

1. Because of the initial state (a material cause).

2. Because the associated tangent vector points in this direction (an efficient cause).

3. Because the tangent vector has this magnitude (a formal cause).

This is the very essence of mechanism.

The *identity* or (intrinsic) nature of the system that is doing the behaving is tied to the *formal causes* of its behaviors—its inertial characteristics, how it sees forces. It is *not* tied to the behaviors themselves in any direct fashion, although it is the behaviors (states and changes of state) that we actually see. It is thus natural to call these identity-determining aspects *genome,* and the behaviors themselves, of which genome constitutes formal cause, *phenotypes.* In Weissmannian terms, phenotypes constitute *soma;* their formal causes are *germ-plasm.*

The forces themselves are the efficient causes of behaviors. The *internal forces* are the efficient causes of behaviors of the system in isolation—the behaviors in an empty environment. This concept would be vacuous unless "parts" of our system could *exert* forces on other parts, as well as respond to them inertially.

Note that the identity of a system, which we have tied to its inertial aspects, has a dual in what we may call its *gravitational* aspects—its capacity to exert forces. We measure inertial properties of a system by looking at how it responds to forces; we measure gravitational ones by seeing how it affects behavior in other systems. At some sufficiently deep level, mechanism has to mandate that inertial and gravitational parameters must coincide (e.g., inertial mass must *equal* gravitational mass).

The main thing to note is the requirement that *behavior cannot affect identity* in an isolated system. A behavior cannot act back on its formal causes. Or, stated another way, an isolated system cannot change its own identity.

Now let us suppose the environment of our system is not empty. That is, we cannot answer why questions about system behaviors in terms of internal forces alone.

We shall call an environmental force *admissible* if it, like the internal forces, does not change the system's identity. That is, its only effect on the system is a change in *behavior*. A nonadmissible force, one that can change the system *identity*, would make that very identity *context-dependent*—make it depend on a larger system for its determination. In particular, reductionism in general rests on a presumed context-independence of identity—namely, that the parts of a larger system are *the same*, either in situ or in vacuo.

Clearly, the change in behavior arising in a system from an external, environmental source (i.e., the discrepancy between actual behavior in a nonempty environment and the corresponding behavior in vacuo) constitutes *forced* behavior. In its turn, forced behavior is distinguished into two classes, conventionally called *autonomous* and *nonautonomous*. In the former case, the tangent vector depends on state alone in its direction; in the latter case, it also depends on time. Autonomous systems must not be confused with closed systems, nor autonomous forces with internal forces.

Nonautonomous admissible forcings are often called *inputs*, and the resultant system behaviors *outputs*. The aim in studying them is to express output as a function (transfer function) of input. In autonomous situations, where time is not involved in characterizing the environment, the goal is to express behavior as a function of time alone.

The restriction to *admissible* forcings, which show up only in system behavior and never in system identity, is crucial to all these kinds of studies. That restriction on *environment* becomes more and more severe as we deal with larger and larger *systems*. Moreover, only admissible environments, which keep identity context-independent, allow us to do traditional mechanics at all. Much of what I shall say henceforth, particularly my comments on survival, pertain to *what happens in nonadmissible environments*.

Ontological Aspects

Epistemology is concerned with the causal underpinnings of system behavior—that is, with unraveling why it changes state the way it does. Ontology, on the other hand, is concerned with existence and creation, hence with why a given system has come to have the identity it does, and the states it does, instead of something else.

Thus when we describe chimera as a new individual, with its own new identity, and its own behaviors (i.e., as a system in its own right), we are asking inherently ontological questions. In particular, we are focusing on chimera as an adaptive response of other systems, other individuals, with their own behaviors and epistemologies.

In general terms, it is simply presumed that ontological questions can merely be *reduced* to epistemological ones. Roughly, the argument is that the "creation" of a new individual is simply another way of talking about the *behaviors of some larger system*—perhaps the entire material universe. Stated another way, any question relating to the ontology of something finds its answers in how that *bigger* system is behaving—that is, how that bigger system is merely changing state.

On the other hand, the need to go to a *larger* system, rather than to smaller ones, to answer ontological questions, as we try to do in all analytical approaches to behavior (epistemology) in systems already given, is a little jarring, to say the least. If nothing else, it requires us to give up the view that *analysis* (as a tool for understanding a given system's behavior in terms of some of its subsystems) and *synthesis* (the assembly or reassembly of such analytic parts, to re-create the system itself) are generally inverse processes (Rosen 1988b). Indeed, we must view *synthesis* as highly *context dependent,* quite in contrast to the completely context-independent (or objective) analytical units that reductionism seeks.

It is not my intention to discuss these matters in depth here. However, they arise naturally in the course of our present discussion; cooperation and competition are concepts that straddle the boundary between epistemology and ontology, analysis and synthesis, behavior and creation. As such, they challenge both.

Survival

To qualify as adaptive, a particular system behavior must be referred to a specific individual, whose survival is favored thereby. Thus the same *behavior,* the same set of objective circumstances, can be adaptive or not, contingent on the individual to which it is referred. On the other hand, though behavior is a mechanistic concept, survival is not; it is uniquely biological.

Intuitively, we see that survival must pertain to the *identity* of a system, which we have tied to the formal causes of its behaviors. That is, survival is tied to the system's array of inertial and gravitational parameters—to what we have called its *genome.* As we have seen above, *admissible* environments, admissible forcings, do not affect such formal causes; they show up rather as efficient causes of behaviors. Hence, survival must inherently involve *nonadmissible* environments—environments that can change system identity. More precisely, it must involve the interplay between system *behavior* and the effect of such a nonadmissible environment on the identity of something. This fact alone is, as we have seen, sufficient to remove survival from the realm of mechanism, in which only admissible forces are allowed.

What behavior can always do, however, is to impact on the environment itself. Indeed, recall that even the most mechanistic system is not only inertial (i.e., it not only responds to forces) but gravitational (i.e., it generates forces), in a way that depends directly on *genome* or identity. These gravitational effects of system on environment are difficult to discuss in conventional mechanistic settings, because typically the environment is assigned no states, and hence does not even behave in the traditional sense; it is a gravitational *source,* not an inertial sink.

Nevertheless, the capacity of system to change environment is the crux of our characterization of behavior as adaptive, and adaptivity as favoring survival. I shall use a language I developed for somewhat different purposes, but which turns out to be appropriate in this context (Rosen 1979).

Very generally, we want to express an *effect* (a behavior, or change of state, which we shall locally call $\dot{x}$) as a function of its *causes.* Conventionally, these causes are three: an initial state x^0 (material cause), identity or genome α (formal cause), and a total force F (efficient cause). Thus we have

$$\dot{x} = \dot{x}(x^0, \alpha, F). \tag{21.1}$$

In conventional mechanics, these *local* relations are given global validity and constitute the equations of motion of the *system*. The total force F may or may not depend on time (i.e., the system may be autonomous or not).

If the forces F are *admissible* (i.e., do not affect identity), this is expressed by the following formal condition:

$$\partial\alpha/\partial F \equiv 0. \tag{21.2}$$

This is the conventional setting for mechanistic environments. On the other hand, in nonadmissible environments, this condition fails; i.e., we must have the following:

$$\partial\alpha/\partial F \not\equiv 0. \tag{21.3}$$

Additionally, we must express the impact of behavior $\dot{x}$ on the environment (i.e., on F). This will be embodied in the formal expression

$$\partial F/\partial\dot{x}. \tag{21.4}$$

In this language, then, a behavior $\dot{x}$ is *adaptive* if its impact on the environment F lowers the impact of the environment on the genome. The condition for adaptivity of $\dot{x}$ is thus

$$\frac{\partial}{\partial\dot{x}}\left(\frac{\partial\alpha}{\partial F}\right) < 0 \tag{21.5}$$

How adaptive the behavior $\dot{x}$ is (i.e., how "fit" the behavior, and by extension the genome that is its formal cause) is clearly measured by how negative condition 21.5 is in the given situation; the fittest behavior is the one for which condition 21.5 is as small as it can get.

Expression 21.2 says that genome is environment dependent, or context independent—that is, there are only admissible forces involved. If expression 21.2 fails (i.e., expression 21.3 holds), the environment F is nonadmissible; genome is itself changing or behaving, and hence there will be a nonzero temporal derivative:

$$d\alpha/dt \neq 0.$$

Hence we may rewrite the adaptivity condition 21.5 for $\dot{x}$ as

$$\frac{\partial}{\partial \dot{x}}\left(\frac{d\alpha}{dt}\right) < 0 \qquad\qquad (21.6)$$

In other words, we may say that $\dot{x}$ is adaptive to the extent that it is an *inhibitor* of genome change in a nonadmissible environment.

I conclude this section with two brief remarks. First, the entire machinery of my earlier analysis (Rosen 1979) may be taken over into the present context of adaptation and survival. Recall that the interpretation I gave to this machinery was "informational," in the sense that the partial derivatives answered questions of the form, If A changes, what happens to B? I interpreted these answers, this "information," in terms of activations and inhibitions, agonisms and antagonisms, for example, and showed that they transmute into a system of ordinary rate equations when, and only when, a corresponding web of differential forms, with these partial derivatives as coefficients, were all *exact*. Otherwise, there were no overarching rate equations at all; the situation becomes *complex* in my sense (see *Life Itself*).

The second observation is that adaptivity, in the sense just described, does involve interactions between behaviors and their genomes—not direct interactions, but indirect ones, through their gravitational effects on environments that are affecting genomes. It is these that constitute the activations/inhibitions, agonisms/antagonisms, etc., that determine adaptivity and fitness. We can thus see, in another way, that considerations of adaptivity require more causal entailment than purely mechanical approaches to systems allow. This extra entailment can, perfectly rigorously, be regarded as the *final causations* that seem inherent in adaptation.

When Is Chimera Formation Adaptive?

In a certain deep sense, the combination of system and environment to constitute a single universe is already an instance of chimerization. But for our purposes, this is not a very useful situation, since environment does not in itself constitute a system in any causal sense; in particular, it does not have in itself a meaningful identity, or genome. Rather, we shall identify a second system in the environment of the first, and instead of looking at the interplay between our initial system and its undifferentiated environment, divide it into the interplay between the two systems and *their* environment.

Our original system had behaviors $\dot{x}$, states x, an identity α, and an environment F. Our new system will, analogously, have behaviors $\dot{y}$, states y, an identity β, and an environment G. Each system is now part of the environment of the other. In general, as always, we will allow our environments to be nonadmissible.

We will allow our systems to interact with each other in the broadest possible way, along the lines indicated in the preceding section. Thus we will say that the systems *compete* when the behaviors of each of them lower the fitness or survivability of the other. We can express this as follows:

$$\frac{\partial}{\partial \dot{x}}\left(\frac{\partial \beta}{\partial G}\right) > 0; \quad \frac{\partial}{\partial \dot{y}}\left(\frac{\partial \alpha}{\partial F}\right) > 0. \tag{21.7}$$

We recall that the behavior $\dot{x}$ is part of the total environment G seen by the second system; likewise, the behavior $\dot{y}$ is part of the total environment F seen by the first. Thus it is not generally true that the most *adaptive* behavior x, with respect to its environment F as a whole, will also be the most competitive—i.e., the one that will make

$$\frac{\partial}{\partial \dot{x}}\left(\frac{\partial \beta}{\partial G}\right)$$

as large as possible. In other words, fitness of a behavior $\dot{x}$ does not generally mean most competitive, measured entirely against another system.

Likewise, we will call an interaction *cooperative* if the following conditions obtain:

$$\frac{\partial}{\partial \dot{x}}\left(\frac{\partial \beta}{\partial G}\right) < 0; \quad \frac{\partial}{\partial \dot{y}}\left(\frac{\partial \alpha}{\partial F}\right) < 0 \tag{21.8}$$

In this case, the behavior of each system increases the survival or fitness of the *other*. It thus *indirectly increases its own fitness* by favoring something that *inhibits* the impact of its total environment on its genome.

This kind of cooperative strategy constitutes, in the broadest sense, a *symbiosis* of our two systems. It is not yet a chimera in our sense, in that it does not yet have a real identity (genome) and behaviors of its own; it is as yet only a kind of direct product of the individual systems that comprise it.

Nevertheless, we can already see that total fitness of a system depends on behaviors $\dot{x}$, $\dot{y}$ satisfying a cooperative condition 21.8 instead of a competitive condition 21.7. Under such circumstances, we can say that cooperation will be favored over competition. Moreover, the *causally* independent behaviors $\dot{x}$, $\dot{y}$ in these situations (i.e., each can be causally understood as *effects* without invoking the other system at all) have become *correlated*. The statement of this correlation looks like what, in mechanics, is called a *constraint*. This observation, in fact, provides the basic clue that provides the next adaptive step from symbiosis to chimera.

Note that we have supposed *no direct interaction* between our two systems, only *indirect* interaction through activations and inhibitions of environmental effects. In other words, the behaviors of each are governed entirely by inertial responses of each to an undifferentiated gravitating environment, and each system's *gravitational* properties are likewise expended entirely into this environment. This indirect interaction results in our correlation of causally independent behaviors of the two systems separately.

But if the systems do interact directly, their joint behaviors are no longer *causally* independent; we can no longer answer *why* questions about the one without invoking the other. A pair of such systems in direct interaction constitutes a new system, with its own behaviors $\dot{z}$, its own states z, its own identity γ, and its own environment H. In terms of the original systems, the *interactions themselves* take the form of constraints, identical relations between behaviors, states, and genomes, out of which $\dot{z}$, z, and γ are built.

In any case, we can now ask about the survival of this new composite system—i.e., about

$$\frac{\partial \gamma}{\partial H},$$

and about the adaptivity of its own behaviors—i.e., about

$$\frac{\partial}{\partial \dot{z}}\left(\frac{\partial \gamma}{\partial H}\right)$$

And finally these must be compared with the corresponding quantifiers for the original free systems. It is easy to write down, in these terms, the conditions under which a pair of interacting systems, in which behaviors are constrained rather than correlated, survives better (i.e., preserves more of *its* genome in a nonadmissible environment) than either system can by itself.

"Survival of the Fittest"

So far, we have couched our discussion entirely in terms of an *individual*, with its own identity (genome), its own behaviors, and its own (generally nonadmissible) environment. We have identified survival with persistence of genome, and measured fitness of a behavior $\dot{x}$ in terms of inhibition of rate of change of genome. Even in these terms, "survival of the fittest" is already highly nontautological.

However, Darwinian evolution does not yet pertain to this view. To say that a behavior x, manifested in an individual *now*, has *evolved* is to raise an ontological question about $\dot{x}$—i.e., about behaviors of behaviors. In a traditional mechanistic sense, we can claim to understand a behavior x completely in traditional Aristotelian causal terms, without involving any larger system. But when we answer the question, Why $\dot{x}$? with a putative answer, Because $\ddot{x}$ (i.e., because of some dynamics on some *space* of behaviors or phenotypes), we are in a different realm. In fact, we are dealing with an entirely new cast of individuals (the ones with $\ddot{x}$ as phenotypes), with their own identities and environments.

Many of the deep puzzles and conundrums of evolutionary theory arise from this opposition. A major factor here is that biologists, as a group, seek simplistic mechanistic explanations for behaviors $\dot{x}$ but do not want corresponding explanations for $\ddot{x}$ (see *Life Itself*).

I have so far omitted from this discussion the primary connection between the individuals that are adapting and the (entirely different) "individuals" that are evolving. That connection rests on a uniquely biological *behavior,* manifested by our adapting individuals, which we may for simplicity call reproduction. In the most elementary terms, this behavior is manifested in the reorganization of environmental materials, along with its own, to generate new individuals with their own identities and behaviors.

The result of this reorganization of the environment, arising from a (gravitational) behavior of our original individual, is to increase the ploidy of its genome and of the behaviors in the new individual, of which identity is formal cause. As with any behavior, we can ask, How adaptive is it? But as we have seen, before we can answer such a question, we must specify the *individual* whose survival is favored by it.

When we begin to address such questions, we find ourselves suddenly transported to quite a different universe, one whose states are now specified by these ploidies, and with how they are changing over a quite different timescale. In this universe, the original adapting individuals are *turning over* (in fact, very rapidly in the new timescale). But behavior in this new universe is just rates of change of ploidies. And this is the universe that must be related to the behavior of behavior, $\ddot{x}$, the ontological universe that I mentioned previously.

In this new universe, the basic concept of survival needs to be completely redefined, to refer to *ploidies* of genomes rather than to the genomes themselves. Symbolically, it now refers to $d\| \alpha \|/dt$, rather than to $d\alpha/dt$, or $\partial\alpha/\partial F$. Accordingly, adaptiveness and fitness must be *redefined* according to whether they activate $d\| \alpha \|/dt$ or not—i.e., according to how they affect ploidy, rather than, as was the case before, how much they protect an individual genome from a nonadmissible environment F. Without going into too many details, there is no reason these two notions of adaptiveness should coincide; they refer to entirely different "individuals."

The longtime behaviors of these ploidies are the province of Darwinian evolution. In particular, *survival* means keeping a ploidy positive. From this viewpoint, the fittest behavior $\dot{x}$ of an *individual* is the one for which

$$\partial/\partial\dot{x}(d\|\alpha\|/dt)$$

is most positive, not the one for which

$$-\partial/\partial\dot{x}(\partial\alpha/\partial F)$$

is. From an evolutionary point of view, these are the fittest, that do the surviving in the new situation.

As always, the nonadmissible environment is regarded as a source for "new genomes" (i.e., those that make a ploidy go from zero to something positive). And with the new genomes come new behaviors. Thus any dynamics on a space of ploidies is reflected in a corresponding dynamics of a phenotypic space of *behaviors*. That dynamics is supposed to generate precisely the $\ddot{x}$—i.e., it is supposed to provide the ontogeny of behaviors manifested by original individual *phenotypes*.

It is traditional in this context to thus measure the fitness of a *genome* α by its effect on its own ploidy—i.e., by

$$\partial\|\alpha\|/\partial\alpha.$$

Competition between two genomes α, β is expressed by the conditions

$$\frac{\partial\|\alpha\|}{\partial\beta} < 0, \; \frac{\partial\|\beta\|}{\partial\alpha} < 0$$

analogous to condition 21.7. The hope is that these quantities suffice to drive an autonomous dynamics expressed locally by the differential relations

$$d\|\dot{\alpha}_i\| = \sum_j \frac{\partial\|\alpha_i\|}{\partial\alpha_j} d\|\alpha_j\|$$

perhaps supplemented by an *admissible* forcing term derived from the originally nonadmissible environment F. And the further hope is that such a dynamics can be projected onto a corresponding behavior space, or phenotype space, where it will assume the form of an action principle.

When Is Chimera the Fittest?

Adaptation is a meaningless concept unless it is tied to an individual whose survival is enhanced by it. Otherwise, it just disappears into dynamics. If we choose our individuals differently (and correspondingly change our idea of what survival means), our notions of what is adaptive will generally change as well.

From the outset, we have tied our identification of an individual to its genome, and hence with the formal causes of its behaviors. A behavior itself could be adaptive or not, in this context, depending on its effect on genome preservation. In this context, then, the fittest behaviors are those that minimize change of genome (identity) in the face of environments that can change it. I have given some very general conditions for the fitness of chimera formation (e.g., cooperation) in this situation.

In general, one of the basic requirements for identifying the kinds of individuals to which such arguments apply is that they themselves must not, by virtue of their own behaviors, be able to change their own genomes. That is, we must not allow their behaviors to act back on their own formal causes. This is true quite generally; in biology, it is the content of the Central Dogma (Judson 1979).

However, behaviors do affect the ploidy of genome, seen in the context of a population. Such a population may itself be regarded as a new individual, whose behaviors can be regarded as adaptive or not. This is the context of *evolutionary* adaptation. In this new context, adaptations are measured by the growth of ploidies.

At heart, the Darwinian concept of survival of the fittest rests on an identification between the two entirely different kinds of fitness of individual behaviors I have just sketched, pertaining to two entirely different kinds of individuals, and hence two entirely different measures of survival and adaptation. Conceptual difficulties with evolution have always grown from the fact that the two need not coincide.

In particular, there is no reason why behaviors that maximize the survival of an individual should also maximize its fecundity, or indeed vice versa. To the extent that these entirely different measures of fitness diverge, to speak of fitness as an abstract property of an individual behavior simply creates an equivocation. Especially so since fecundity arises from individual behaviors that may be very far from adaptive in the sense of individual survival.

Equivocations of this kind spawn apparent paradoxes. A simple example is the so-called Galileo paradox, which involved the "size" (i.e., cardinality) of sets. Galileo asked which is bigger: the set of all integers or the set of even integers. Judged simply by inclusion, and according to the Euclidean axiom that a whole (all the integers) is bigger than any proper part (the even integers), we conclude that the former is clearly bigger. But, as measured by enumeration, we must conclude that the two are the same size. Two entirely different measures of size are involved here: measures that coincide for finite things but that can disagree for infinite ones. Indeed, the Galileo paradox ended, in the hands of Georg Kantor, as the diagnostic property that separated what is finite from what is not.

So it is with concepts such as fitness, and especially so when we attempt to compare fitnesses. Here, we treat a chimera as a new individual and ask about its fitness as such. Moreover, just as in the Galileo paradox, we try to compare its fitness with those of other individuals, especially against those of its constituents, in terms of the advantages of cooperation against noncooperation.

What I have tried to do in the this discussion is rather more modest. I have tried to argue that chimera formation culminates in the generation of new kinds of individuals; it causes new identities, new genomes, and new behaviors to emerge, which could never be generated in any other way, and certainly not by processes of differentiation alone. I have tried to give conditions under which chimera formation is adaptive, in the sense that it favors the survival of its constituents more than the survivals they could achieve otherwise. But I certainly cannot say that, if fitness is measured in terms of ploidies, as it is when we speak of Darwinian evolution, such considerations have any significance at all. The two issues are clearly not the same. Indeed, an establishment of relations between these two distinct playgrounds for adaptiveness, the evolutionary and the physiological, would in itself be an instance of chimera formation.

CHAPTER 22

ᐇ

Are Our Modeling Paradigms Nongeneric?

I have been, and remain, entirely dedicated to the idea that modeling is the essence of science and the habitat of all epistemology. Although I have concentrated my efforts on biology and the nature of organism, I have also asserted for a long time that human systems and organisms are very much alike—i.e., they share, or realize, many common models. I was a pioneer in the wide deployment of mathematical ideas for purposes of modeling, particularly in the area of stability theory; indeed, I wrote perhaps the first modern text devoted to this purpose (1970).

What follows is a natural extension and continuation of these efforts. However, the tactics of modeling, and the mathematical machinery currently regarded as the only way to implement these tactics, are much too narrow for our scientific purposes. This mathematical machinery, and hence what it allows us to capture about the world around us, necessarily misses most of what is really going on.

Let me be a little more specific. Any mathematical structure, any mathematical system, whether it be an explicit model of something in the material world or not, may possess the property of being *formalizable*. There are many ways of describing this property, but they all amount to being able to describe the entire system as software to a mathematical machine (a Turing machine), in such a way that the machine can *simulate* the system. Everything about a formalizable system can be expressed as pure syntax; every inferential process in such a system can be thought of as rote symbol manipulation or word processing.

Concern with formalizability arose historically from the need to eliminate paradoxes and inconsistencies in mathematical operations that had been assumed to be free from them. Hilbert and others argued that every mathematical system was thus formalizable, and hence that mathematics itself can and must be expressed as a game of pattern generation on a set of symbols, devoid of any semantic component. However, Hilbert's program was destroyed by the celebrated Gödel Incompleteness Theorem, which showed that the property of being formalizable was exceedingly special, excessively special among mathematical systems—i.e., that one cannot capture very much of conventional mathematics in purely formalistic, syntactic terms.

The special character of computability or simulability spills over into the sciences, through the intervention of mathematical models to characterize material phenomena. The idea that every model of a material system must be simulable or computable is at least tacitly regarded in most quarters as synonymous with science itself; it is a material version of Church's Thesis (i.e., effective *means* computable) (Rosen 1962a, 1988a).

I call a material system with only computable models a *simple system,* or *mechanism.* A system that is not simple in this sense I call *complex.* A complex system must thus have noncomputable models.

To say that material systems may be complex in this sense and, in particular, to assert that organisms, or human systems, are thus complex, is a radical thing to do. For one thing, it says that differential equations, and systems of differential equations (i.e., dynamical systems), which are inherently simulable, miss most of the reality of a complex system . . . just as any attempt to formalize, for example, Number Theory misses most of its theorems. It does not say that we learn nothing about complex systems from simple models; it merely says we should widen our concept of what models are.

I shall proceed with a discussion of the concept of *genericity,* culminating in an argument that simple systems are nongeneric (rare). I will then discuss the related concept of *stability,* and the testing for stability by applying generic perturbations. I will conclude by showing that dynamical systems, systems of differential equations, become complex when generically perturbed, and I will briefly discuss what this means for the scientific enterprise.

Genericity

In an intuitive mathematical context, genericity means that which is *typical,* that which is devoid of special qualifications or properties superimposed on the typicality in question. The best way to introduce this concept is through some familiar examples.

1. It is generic for a real number to be irrational; it is nongeneric for a number to be rational, or even to be computable in the usual sense.

2. It is generic for two lines in a plane to intersect; it is nongeneric for them to coincide or be parallel.

3. It is generic for a square matrix to be invertible, and hence it is generic for its determinant not to vanish. It is accordingly generic for a set of $\leq N$ vectors in an N-dimensional space to be linearly independent. Linear dependences, vanishing of determinants, and noninvertibility of matrices are thus nongeneric.

4. It is generic for a differential form to be nonexact or nonintegrable.

5. It is generic for sets to be infinite.

Generic properties are thus what we expect to see when we approach something in an objective, unbiased way. We are very strongly biased, for instance, in the direction of rational numbers; this is why irrationalities were so named in the first place. Nevertheless, by any *objective* criterion, it is the rational numbers that are the rare and special ones, and our predilection for them tells more about us than about numbers.

Indeed, recall that rational numbers are those represented by terminating or periodic decimal expansions. This illustrates the proposition that nongenericities are encumbered with special properties, degeneracies, symmetries, which in fact characterize or separate the generic from the special.

Ironically, in mathematics it is often the nongeneric that yields the theorems, precisely because of all the special conditions that define them. This fact is what thrusts nongenericity into the spotlight when modeling material reality, even though we know that, for example, Hamiltonian or conservative systems are nongeneric; that is where we have come to think the theorems are.

Yet the property of being typical, of having no special properties, is itself a property. Arguing from the typical is, for instance, what underlies the Thom Classification Theorem (1976), or in quite a different area, what is behind the notion of forcing, by means of which Paul Cohen (1966) could prove the independence of the continuum hypothesis (and the axiom of choice) from the rest of Set Theory. It expresses the robustness so often invoked to justify conclusions drawn from a model or a metaphor.

What is the diagnostic of genericity? What is the difference between being typical and being special? As Thom pointed out, it is this: An *arbitrary* (i.e., a *generic*) perturbation of what is generic yields something again generic; an arbitrary perturbation of what is nongeneric destroys the special properties (e.g., breaks the symmetries, or lifts the degeneracies) on which nongenericity depends.

There is a little circularity in this intuitive treatment: namely, we need to characterize what a generic *perturbation* is before we can decide, by this criterion, whether what we perturb is generic or not. If we do not address this properly, and we tacitly mandate perturbations that are themselves nongeneric, we will get wrong answers. This, as we shall see, is the crux when trying to decide whether computability or simulability of something is nongeneric.

The diagnostic of comparing a perturbed and an unperturbed situation is what relates genericity to ideas of stability and bifurcation. But they are not the same, and it is important to distinguish the province of stability from that of genericity. Thus we must turn now to a discussion of what they have in common and what they do not.

Stability

Stability also requires us to compare a perturbed with an unperturbed situation. But central to stability questions is the requirement that our perturbations be in some sense *small* (or *small enough*)—i.e., that there be an idea of *metric approximation*. We must determine whether a first-order change in some system characteristic produces a higher-order change in another (stability), or a lower-order change (instability), or neither.

The prototypic situation for stability studies involves playing off an equivalence relation (i.e., a notion of similarity) against a metric (Rosen 1970). The equivalence relation partitions the metric space into a family of disjoint equivalence classes. A point is *stable* if it lies in the interior of its equivalence class; there is thus a whole neighborhood, a sphere of some finite radius, about a stable point, which lies entirely in the class. Thus a point is stable if every sufficiently good approximation to it is also similar to it.

A point that is not stable is called a *bifurcation point.* Any neighborhood of a bifurcation point must intersect more than one equivalence class; that is, no matter how closely we metrically approximate to such a point, that does not guarantee similarity.

We can ask a question like the following: Is it generic for a point to be stable (i.e., are bifurcation points nongeneric)? We might be tempted to answer yes, arguing that bifurcation points must lie on lower dimensional boundaries separating open equivalence classes. But why should there be any open equivalence classes at all? Presuming that, in advance, is itself a bias, such as the one we have for rational numbers, and a nongeneric one at that. And if there are none, then every point is in fact a bifurcation point. Indeed, in general, stability will be neither generic nor nongeneric.

We are concerned with one particular aspect of bifurcation. Namely, what happens when we *restrict* the perturbations we apply to a point, so that we no longer have generic perturbations available to us? In this case, we clearly cannot reach every point in an open neighborhood, so we cannot be sure we can sample every nearby equivalence class. In more formal terms, we no longer can be sure that an *unfolding* of a bifurcation point is a versal or *universal unfolding.* It is by precisely such undue restriction on perturbations that complexity is concealed.

Some Basic Terminology

To discuss ideas of genericity directed toward dynamical systems (e.g., systems of differential equations) and their perturbation, with an eye to their ultimate role as *models* of real-world phenomena, it is necessary to introduce some basic concepts and the terminology that describes them.

Let X be a manifold, something that looks like a chunk of a Euclidean space of some finite dimension. In particular, X is thus a metric space. Moreover, we can differentiate certain functions defined on X; it suffices to consider real-valued functions. We will think of X as a prototypic *state space*. However, our main area of interest for the moment is not X, but functions defined *on* X. Let us denote these functions by $H(X, Y)$, where Y is another set. Since we are considering real-valued functions on X, we can fix the range Y of our functions as the real numbers $\mathbb{R}$.

Now $H(X, \mathbb{R})$ is not in general a manifold; it does not look like a piece of Euclidean space. However, it can be turned into a metric space in many ways; any such metric allows us to discuss *approximations* of a function f in $H(X, \mathbb{R})$ by others.

We usually write such a function from X to $\mathbb{R}$ in the notation

$$f(x_1, \ldots, x_n),$$

which we can do because X is a manifold; thus it must have (local) coordinate systems. In this notation, the arguments x_i, the coordinates of a point in the manifold, are interpreted as *state variables* of any material system that X models.

But this notation is incomplete. It omits or conceals tacit arguments of f, which are usually called *parameters*. These can be thought of as numbers whose values must be specified before we can evaluate f at a state. Hence, in some way, these numbers determine or specify f itself. If there are r such numbers, say $a_1, \ldots, a_r$, then the value of a function f at a state is really determined by $n + r$ arguments, which can be somewhat redundantly expressed as

$$f(x_1, \ldots, x_n,\ a_1, \ldots, a_r).$$

In this notation, there is no mathematical distinction between the state values and the parameters; they are all equally arguments of the function f. However, a slight change in notation will completely change this. Let us write

$$f(x_1, \ldots, x_n,\ a_1, \ldots, a_n) = f_{a_1 \ldots a_r}(x_1, \ldots, x_n).$$

The effect of this is to change a single function of $n+r$ arguments into a parameterized family of functions of only n arguments. Thus the parameters now appear as local coordinates *in the function space* $H(X, \mathbb{R})$. We can ask what happens in this parameterized family as (1) we vary the state variables $x_1, \ldots, x_n$, or (2) we change the coordinates or parameters $a_1, \ldots, a_r$. And, most important, we can ask whether either (1) or (2) is a *generic* way of exploring a whole neighborhood of a given function f in $H(X, \mathbb{R})$.

We have just seen that we can turn a *fixed* function of $n+r$ arguments into an r-parameter *family* of functions of n arguments. It is clear that this procedure works the other way: *an r-parameter family of functions of n arguments can be turned into a single fixed function of* $n+r$ *arguments.* This is the result we shall need.

In the light of these ideas, let us consider the following particular way of introducing parameters. Suppose we have a function of n arguments, say $f(x_1, \ldots, x_n)$. Let us take N more functions of these same arguments, say

$$g_i(x_1, \ldots, x_n), \quad i = 1, \ldots, N,$$

and use these to generate new functions of the form

$$f + \sum_{i=1}^{N} \epsilon_i g_i. \tag{22.1}$$

Clearly, if the numbers ϵ_i are small, then 22.1 can be regarded as a perturbation of f, or an approximation to f. The totality of these constitute what is called an *unfolding* of f, for reasons we shall come to in a moment.

On the other hand, there is a fixed function $F = F(x_1, \ldots, x_n, \epsilon_1, \ldots, \epsilon_N)$ of $n+N$ arguments, such that

$$F = f + \sum_{i=1}^{N} \epsilon_i g_i. \tag{22.2}$$

Clearly, $F = f$ when the parameter values ϵ_i are put equal to zero. Moreover, a variation of the *function* f via the previously described unfolding is equivalent to a variation of the *arguments* of the fixed function F.

If we think of a function as constituting *hardware,* and its arguments as *software,* then the concept of unfolding says that there exists hardware (the function *F*) that can be varied only through manipulations of its arguments (i.e., of its software). What looks like an *arbitrary* perturbation of the original function *f,* via the unfolding, becomes *equivalent to a very special perturbation of the function F.* And conversely, a truly arbitrary (generic) perturbation of *F* does something quite drastic around *f.*

The existence of fixed hardware, which communicates with its ambient environment only via its software, is one of the essential features of computability or simulability—the essence of a Turing machine and hence of pure syntax. The direct action of an environment on the hardware itself is strictly, if tacitly, forbidden. And this in turn is the essence of a simple system—that we be able to reach a function such as *F,* completely closed off from its environment (and hence from its neighbors) only through variations of its arguments.

Dynamics

So far, I have talked only about metric aspects—i.e., about perturbations or approximations of functions on a manifold *X,* and about parameterized families of such functions (e.g., unfoldings). To talk about stability, and in particular about stable parameterized families, we need an equivalence relation on $H(X, \mathbb{R})$. We can get one by turning our functions, and their parameterized families, into dynamical systems.

We can go from functions to dynamical systems by the deceptively simple formal device of identifying the value of a function at a point x of our manifold, with a tangent vector at x—i.e., by putting

$$d\mathrm{x}/dt \;=\; f(\mathrm{x}). \tag{22.3}$$

Remembering that *f* contains parameters, which in fact determine what its values will be, we can see that the space of these parameter values maps into the tangent spaces at each state x; that is why parameters are often called *controls.*

Call two dynamical systems *equivalent* if there is a coordinate transformation on the manifold $\mathbb{X}$ that maps the trajectories of the first onto the trajectories of the second. If one of these dynamical systems is ob-

tained by changing the parameter values of the other, then their equivalence means that we can wipe out the effect of that parameter variation
by means of a compensating coordinate transformation in $\mathbb{X}$ alone. It is
easy to see that this is a true equivalence relation on dynamical systems
on $\mathbb{X}$ (reflexive, symmetric, and transitive), and hence on the functions
on $\mathbb{X}$ that define them via 22.3 above. It is with respect to this relation
that we can talk about stability, about the interplay between this equivalence relation and the metric on $H(X, \mathbb{R})$.

More specifically, let us look at a parameterized family $f_\alpha(\mathbb{x})$ of such
functions. This gives a parameterized family of dynamical systems

$$d\mathbb{x}/dt \;=\; f_\alpha(\mathbb{x}). \qquad (22.4)$$

We can now ask whether this is a *stable* family—i.e., lying within a
single equivalence class. In turn, this means that we can go from the
trajectories of any system in the family to any other by a change of
coordinates in X, which depends *only* on the parameter values that specify the two systems in the family.

A parameter vector α^* is thus called *stable* if it has a neighborhood
U such that all the systems f_α, $\alpha \in U$, constitute a stable family. Otherwise, we call α^* a *bifurcation point*. Clearly, if α^* is a bifurcation point,
then no matter how small we choose the neighborhood U, there will be
a parameter vector $\alpha \in U$ such that f_α and f_α^* are dissimilar: their trajectories are not intertransformable by a mere coordinate transformation of
X. Bifurcation here thus means that there are perturbations $\alpha^* \mapsto \alpha$,
metrically as small as we please, that cannot be undone by just a change
of coordinates. Any such U thus intersects several similarity classes.

Next, let us suppose that the parameterized family 22.4 comes from
an unfolding, something of the form 22.2. Thus the parameter vector
we have called α should now be called ϵ. The question is now, If ϵ^* is a
bifurcation point, and if U is a small neighborhood of ϵ^*, how many of
the similarity classes near f_ϵ^* can actually be reached by perturbing ϵ^*?
This is a highly nontrivial question because these similarity classes are
independent of any parameterizations and any unfoldings; they depend
only on the metric in $H(X, \mathbb{R})$ and on the equivalence relation or similarity.

An unfolding is called *versal* if a neighborhood U of a parameter
vector ϵ^* in 22.3, which is only an N-dimensional thing, is nevertheless

big enough to touch every equivalence class near f_ϵ^*. The unfolding is called *universal* if the dimension N of U is as small as possible: that is, any lower-dimensional U will miss some of the similarity classes near f_ϵ^*.

The habitat of catastrophe theory constitutes such unfoldings for those very special functions f that admit versal and universal parameterized families. The Thom Classification Theorem proceeds by examining what happens generically in the neighborhood of these very special functions and their unfoldings. We shall not be concerned with this, but rather with the nongenericity of the concept of unfolding itself. The root of this resides in the idea that there is ultimately a single fixed function F that generates the unfolding, and that can be varied only through variation of its arguments.

As I have said, nongenericity seems to generate the theorems, and the Classification Theorem is a case in point. Where its hypotheses hold, we can learn many deep and important things. But those hypotheses already restrict us to the realm of *simple systems,* and to perturbations that keep us restricted to this realm.

Dynamics and "Information": AI Networks

Given a dynamical system

$$d\mathsf{x}/dt \;=\; f(\mathsf{x}), \text{ or } dx_1/dt \;=\; f_i(\mathsf{x}),\tag{22.5}$$

where we have for the moment omitted writing any explicit parameters, we can associate with it a variety of other quantities that express important aspects of its local behavior.

For instance, following Higgins (1967), we can introduce the quantities

$$u_{ij}(\mathsf{x}) \;=\; \partial/\partial x_j (dx_i/dt)$$
$$= \; \partial f_i/\partial x_j.$$

These state functions u_{ij} have the significance that their signs at a state x tell how a change in a state variable x_j is translated by the dynamics into a corresponding effect on the rate of change of another, x_i, in that state x.

More specifically, let us call x_j an *activator* of x_i at a state x if

$$u_{ij}(\text{x}) \; > \; 0.$$

Intuitively, an increase in x_j at x will be reflected in an increase in the rate of production dx_i/dt of x_i; likewise, a decrease in x_j will decrease that rate of production. On the other hand, if

$$u_{ij}(\text{x}) \; < \; 0,$$

then we can call x_j an *inhibitor* of x_i in that state. The matrix of functions u_{ij} constitutes what I call an *activation-inhibition network* (Rosen 1979).

The original interest in these networks resides in the fact that *activation* and *inhibition* are informational terms, pertaining to semantic aspects of the dynamical behavior. There are many situations (e.g., in brain theory, or in ecology) where it seems more natural to consider the AI network as primary. So we have the question, Given such a network, can we always go back to a generating dynamics, a set of rate equations such as 22.5, that generate them?

Writing such rate equations means in effect producing the functions f_i. The above argument produces not the f_i themselves, but rather a set of relations that the *differentials df_i* must satisfy—namely

$$df_i \; = \; \sum_{j=1}^{n} u_{ij}\, dx_j \tag{22.6}$$

If there is to be a set of rate equations that produce a given network $[u_{ij}(\text{x})]$, then, *the differential forms 22.6 must all be exact.*

But it gets worse. We can iterate the ideas that led to the AI patterns as follows. Let us introduce the quantities

$$u_{ijk}(\text{x}) \; = \; \partial u_{ij}/\partial x_k.$$

The signs of *these* quantities reflect the effects of a change in a state variable x_k on the activation or inhibition which x_j imposes on x_i. Thus if

$$u_{ijk}(x) > 0$$

in a state, we may call x_k an *agonist* of the activation (or inhibition) of x_i by x_j—otherwise, an *antagonist*. These are again semantic, informational terms, which are defined whenever we start from a set of rate equations, but which seem independently meaningful.

In any case, we must superimpose the pattern of agonism/antagonism, embodied in the functions u_{ijk}, on the AI pattern obtained previously. And of course, we can iterate this process indefinitely.

At each step of this process, we obtain another collection of differential forms such as 22.6. Again, if we start from a system of rate equations such as 22.5, they are all exact. Moreover, we can pass from any informational level to any other, and from any of them to the generating rate equations, just by differentiating and/or integrating. Thus in this case, all these layers of informational or semantic interactions are intertransformable and equivalent—just another way of talking about rate equations.

But surely, if we generically perturb a nongeneric thing such as an exact differential form, we will get a nonexact (generic) form. So suppose we try to unfold the functions f_i in 22.5—i.e., replace 22.5 by a system of the form

$$dx_i/dt = f_i + \sum_{k=1}^{n} \epsilon_k g_k \qquad (22.7)$$

where the ϵ_k are small parameters and the g_k are arbitrary but fixed functions. What happens to the u_{ij} and, in particular, to the differential forms 22.6 that they define?

It can be immediately verified that the new AI pattern $[u'_{ij}(x)]$, coming from 22.7, *still gives us an exact differential form*. That is, what we thought of as a *generic* perturbation of the rate equations translates into a highly nongeneric perturbation of the associated AI pattern (and also, of course, of the agonist/antagonist pattern, and all of the iterates of this process).

Conversely, a *generic* perturbation of the AI pattern, which will necessarily make the differential form 22.6 *inexact*, translates into something very peculiar in the vicinity of the original rate functions f_i.

Complex Systems

I shall now touch on some ramifications of the preceding, especially in the modeling of material systems. As we have just seen, what looks like a generic perturbation of a system of rate equations (i.e., a dynamical system) necessarily gives rise to a highly nongeneric perturbation of the infinite networks of differential forms that coincide with the dynamics (e.g., the AI pattern, and the agonist/antagonist pattern). On the other hand, a generic perturbation of the AI pattern, or any of the others, renders all these patterns independent and wipes out the concept of rate equations entirely. It thus looks like there is something very special, very degenerate, about systems of rate equations, and about the ways we have been perturbing them.

In a nutshell, what is special is precisely what we drew attention to before: the tacit presupposition that everything can be described in terms of one overarching function (we called it F above) that can be varied only through variation of its arguments—i.e., variations in states and parameters. When we know this F, we know everything about the system there is to know; in particular, every *model* of the system can be obtained from it, through restrictions on its arguments alone. Indeed, F thus appears as the *largest* model.

This property of possessing a largest model is a hallmark of simple systems or mechanisms. It is basically what allows a whole system to be expressible as software (i.e., as program and data) to an extraneous symbol-processor or mathematical machine, and hence to make the system simulable.

The restriction that a function F cannot be varied except through variation of its arguments, which we may express formally by mandating that

$$(\delta F)(x) = F(\delta x),$$

places a terribly strong restriction on what can be in a *neighborhood* of F. Conversely, if we try to vary F *only* through variations δx of its arguments, we cannot thereby see what is typical (generic) in such a neighborhood, because we are trying to explore that neighborhood through means that are themselves highly nongenetic.

In the preceding section, I sketched a way to lift this nongenericity.

Instead of trying to perturb a system of rate equations in the usual way, we apply a perturbation to one or more of the systems of exact differential forms that are ordinarily completely equivalent to the system itself, and to each other. That decouples everything, very much like the splitting of a degenerate spectral line when a perturbing magnetic field is applied (Zeeman effect).

It should be intuitively clear that such a decoupling renders the system *complex*. It is no longer possible to express what happens in it in the form of a finite program to a syntactic simulator. It is further clear that complexity is thus *generic* in the usual sense; perturbing a complex system as we have done gives us another complex system, while perturbing a simple one does not preserve simplicity.

Some Implications for Modeling

There are many deep ramifications of these ideas, which I can only hint at in this short space, but I can touch on a few.

In *Anticipatory Systems,* I characterized the modeling relation as a kind of congruence between two systems of entailment. The mathematical model is a system of inferential entailment put into such a congruence with a real-world system of causal entailment. The property of simulability in a model of something turns out to impose very strong limitations on causality in what it models. In fact, simulability mandates an extreme degree of causal impoverishment. Almost everything about such a system is unentailed from within the system and must accordingly be externally *posited;* all that remains is the entailment of "next state" from "present state." It is precisely this paucity of entailment in simple systems or mechanisms that allows them to be expressible as software.

Complex systems may be thought of, on the other hand, as manifesting more entailment (more causality) than can be accommodated by a mechanism. Organisms, for example, sit at the other end of the entailment spectrum from mechanisms; almost everything about them is in some sense entailed from something else about them.

On the other hand, in a complex system, there is no meaningful intrinsic distinction between hardware and software, no single overarching function that stays fixed while only its arguments can vary. In material terms, a system of this type is literally infinitely open, whereas a mechanism or simple system can be, at best, finitely open. The upshot of this is that if we try to replace a complex system by a simple one, we necessarily miss most of the interactions of which it is capable. Herein lies the primary basis for the counterintuitive characteristics or nonmechanistic behavior so often manifested by organisms; the causal entailments on which they depend are simply not encoded into the simulable models we are using.

No superposition of simple models will yield a complex system; we cannot leave the realm of computability in this fashion, any more than we can build an infinite set by means of finite operations on finite sets. Thus, in general, it is not a good tactic to try to study open systems by opening closed ones; it usually turns out that closure, in the material sense, is so degenerate (i.e., so nongeneric) that the behavior of a perturbed closed system will depend much more on how it was perturbed than on its behavior when closed.

It must be emphasized that we can still make dynamical models of complex systems, just as we can formalize fragments of Number Theory. We can approximate, but only locally and temporarily, to inexact differential forms with exact ones under certain conditions. But we will have to keep shifting from model to model, as the causal structure in the complex system outstrips what is coded into any particular dynamics. The situation is analogous to trying to use pieces of planar maps to navigate on the surface of a sphere.

Indeed, just as a sphere is in some sense the limits of its approximating planar maps, a complex system can be regarded as a *limit* of simple ones. Pursuit of this analogy, however, would take us far beyond the scope of this chapter.

I will conclude by remarking that concepts such as activation/inhibition and agonism/antagonism, which are informational (semantic) terms, may be used to introduce a language of function into the study of (complex) systems. Here, I use the word *function* in the biological rather than the mathematical sense—e.g., the function of X is to do Y. For instance, in the dynamical example described in the Dynamics section, we could identify the function of a state variable x_i in a state x with what it activates and inhibits in that state. We thus inch toward a legitimation of the Aristotelian category of final causation, bound up with what something entails rather than with what entails it. In complex systems, it is not only completely legitimate to use such a language, it is absolutely necessary. Indeed, this is another fundamental way in which complexity differs from mechanism. Using this kind of language leads us in the direction of *relational models,* which have proved most appropriate for biological purposes (and, by implication, for any kind of human or social system).

References

Bellman, R. 1957. *Dynamic Programming*. Princeton, N.J.: Princeton University Press.

——. 1961. *Adaptive Control Processes*. Princeton, N.J.: Princeton University Press.

——. 1968. *Some Vistas of Modern Mathematics*. Lexington, Ky.: University of Kentucky Press.

Bellman, R., and S. Dreyfuss. 1962. *Applied Dynamic Programming*. Princeton, N.J.: Princeton University Press.

Bellman, R., J. Jacquez, and R. Kalaba. 1960. Some mathematical aspects of chemotherapy: I. One-organ models. *Bulletin of Mathematical Biophysics* 22: 181–98.

Bergman, H. 1973. The controversy concerning the laws of causality in contemporary physics. In Cohen, R. S., and Wartowsky, M. W., eds. *Logical and Epistemological Studies in Contemporary Physics*. Dordrecht: Reidel, 395–462. (*Boston Studies in Philosophy of Science,* vol. XIII)

Beyerchen, A. D. 1977. *Scientists under Hitler: Politics and the Physics Community in the Third Reich*. New Haven, Conn.: Yale University Press.

Bierce, A. 1911. *Devil's Dictionary.*

Clark, R. W. 1972. *Einstein: The Life and Times*. New York: Avon.

Cohen, P. J. 1966. *Set Theory and the Continuum Hypothesis*. Reading, Mass.: W. A. Benjamin.

Danielli, J. F. 1974. Genetic engineering and life synthesis: An introduction to the review by R. Widdus and C. Ault. *International Review of Cytology* 38: 1–5.

Davis, M. 1958. *Computability and Unsolvability*. New York: McGraw-Hill.

Dyson, F. 1979. *Disturbing the Universe*. New York: Harper and Row.

Eddington, A. S. 1946. *Fundamental Theory*. Cambridge: Cambridge University Press.

Einstein, A. 1933. *Origins of the General Theory of Relativity.* Lecture of the George

A. Gibson Foundation in the University of Glasgow, June 20. Glasgow: Glasgow University Publications, no. 30.

Einstein, A., and L. Infeld. 1938. *The Evolution of Physics*. New York: Simon and Schuster.

Elsasser, W. M. 1958. *The Physical Foundation of Biology*. New York: Pergamon Press.

De Groot, S. R., and P. Mazur. 1963. *Non-equilibrium Thermodynamics*. New York: Elsevier.

d'Espagnat, B. 1976. *Conceptual Foundations of Quantum Mechanics*. Reading, Mass.: W. A. Benjamin.

Gelfand, Y., and S. V. Fomin. 1962. *Calculus of Variations*. New York: Prentice Hall.

Glasstone, S., K. J. Laidler, and H. Eyring. 1941. *The Theory of Rate Processes*. New York: McGraw-Hill.

Gödel, K. 1931. Über formal unentscheidbare Sätze der Principia Mathematica und verwandter Systeme. *Montshefte für Mathematik und Physik* 38: 173–98.

Guthrie, W. K. C. 1962. *A History of Greek Philosophy*, vol. 1. Cambridge: Cambridge University Press.

Hadamard, J. 1923. *Lectures on Cauchy's Problem in Linear Partial Differential Equations*. New Haven, Conn.: Yale University Press.

Haken, H. 1977. *Synergetics*. Berlin: Springer-Verlag.

Halmos, P. R. 1950. *Measure Theory*. New York: van Nostrand.

Handler, P. Introduction. In Handler, P., ed. 1970. *Biology and the Future of Man*. Oxford: Oxford University Press, 3–6.

Hardin, G., and J. Baden. 1977. *Managing the Commons*. San Francisco: W. H. Freeman.

Harmon, L. D., and E. R. Lewis. 1966. Neural modeling. *Physiological Reviews* 46: 513–91.

Higgins, J. 1967. Oscillating reactions. *Industrial and Engineering Chemistry* 59: 18–62.

Jacob, F., and J. Monod. 1961. On the regulation of gene activity. *Cold Spring Harbor Symposia on Quantitative Biology* 26: 193–211.

Judson, H. F. 1979. *The Eighth Day of Creation*. New York: Simon and Schuster.

Kleene, S. C. 1950. *Introduction to Metamathematics*. Princeton, N.J.: van Nostrand.

Langton, C. G., ed. 1989. *Artificial Life*. New York: Addison-Wesley.

Langton, C. G., C. Taylor, and J. D. Farmer, eds. 1992. *Artificial Life*, II. New York: Addison-Wesley.

Löfgren, L. 1968. An axiomatic explanation of complete self-reproduction. *Bulletin of Mathematical Biophysics* 30: 415–29.

Mandelbrot, B. 1977. *Fractals, Form and Chance.* San Francisco: W. H. Freeman.

Maurin, K. 1997. *The Riemann Legacy: Riemannian Ideas in Mathematics and Physics.* Boston: Kluwer Academic.

McCulloch, W., and W. Pitts. 1943. A logical calculus of the ideas immanent in nervous activity. *Bulletin of Mathematical Biophysics* 5: 115–33.

McLaren, A. 1976. *Mammalian Chimaeras.* Cambridge: Cambridge University Press.

Mickens, R., ed. 1990. *Mathematics and Science.* Singapore: World Scientific.

Monod, J. 1971. *Chance and Neccessity.* (Trans. A. Wainhouse.) New York: Knopf.

Nicolis, G., and I. Prigogine. 1977. *Self-organization in Nonequilibrium Systems.* New York: Wiley.

Pais, A. 1982. *Subtle Is the Lord.* New York: Oxford University Press.

———. 1991 *Niels Bohr's Times.* Oxford: Clarendon Press.

Pontryagin, L. S., V. G. Boltyanskii, R. V. Gamkrelidze, and E. P. Mishchenko. 1961. *The Mathematical Theory of Optimal Processes.* New York: John Wiley and Sons.

Poston, T., and I. N. Stewart. 1978. *Catastrophe Theory and Its Applications.* London: Pittman.

Prigogine, I. 1947. *Étude Thermodynamique des Phénomènes Irréversibles.* Liège: Desoer.

———. 1967. *Introduction to Thermodynamics of Irreversible Processes.* New York: Wiley.

Ramón y Cajal, S., and F. de Castro. 1933. *Elementos de Técnica Micrográfica del Sistema Nervioso.* Madrid: Tipografía Artística.

Rashevsky, N. 1960. *Mathematical Biophysics: Physico-Mathematical Foundations of Biology,* 3rd ed., vols. 1, 2. New York: Dover.

Rosen, R. 1959. On a logical paradox implicit in the notion of a self-reproducing automata. *Bulletin of Mathematical Biophysics* 21: 387–94.

———. 1962a. Church's thesis and its relation to the concept of realizability in biology and physics. *Bulletin of Mathematical Biophysics* 24: 375–93.

———. 1962b. The derivation of D'Arcy Thompson's theory from the theory of optimal design. *Bulletin of Mathematical Biophysics* 24: 279–90.

———. 1964. The Gibbs paradox and the distinguishability of physical systems. *Philosophy of Science* 31: 232–36.

———. 1967. *Optimality Principles in Biology.* New York: Plenum Press.

———. 1968. Turing's morphogens, two-factor systems and active transport. *Bulletin of Mathematical Biophysics* 30: 493–9.

——. 1970. *Dynamical Systems Theory in Biology.* New York: Wiley Interscience.

——. 1972. Some relational cell models: The metabolism-repair system. In Rosen, R., ed. *Foundations of Mathematical Biology,* vol. 2. New York: Academic Press, 217–53.

——. 1974. Planning, management, policies and strategies: Four fuzzy concepts. CSDI paper. *Quarterly Bulletin of Theoretical Biology* 7(1), and *International Journal of General Systems* 1: 245–52.

——. 1975. Biological systems as paradigms for adaptation. In Day, R. H., and T. Groves, eds. *Adaptive Economic Models,* 39–72. New York: Academic Press.

——. 1977. Book review. René Thom: Structural stability and morphogenesis. (1976; trans. D. H. Fowler.) *Bulletin of Mathematical Biophysics* 39: 629–632.

——. 1978a. Dynamical similarity and the theory of biological transformations. *Bulletin of Mathematical Biophysics* 40: 549–79.

——. 1978b. *Fundamentals of Measurement and the Representation of Natural Systems.* New York: Elsevier.

——. 1978c. Cells and senescence. *International Review of Cytology* 54: 161–91.

——. 1978d. Feedforwards and global systems failure: A general mechanism for senescence. *Journal of Theoretical Biology* 74: 579–90.

——. 1979. Some comments on activation and inhibition. *Bulletin of Mathematical Biophysics* 41: 427–45.

——. 1981. Morphogenesis in networks. In *Progress in Theoretical Biology,* vol. 6. New York: Academic Press.

——. 1983. The role of similarity principles in data extrapolation. *American Journal of Physiology* 224: 591–9.

——. 1984a. Genomic control of global features in morphogenesis. Presented at *Heidelberg Workshop on Pattern Generation,* June 1983. In *Modelling of Patterns in Space and Time. Lecture Notes in Biomathematics* 55: 318–30. Berlin: Springer-Verlag.

——. 1984b. Order and disorder in biological control systems. In Lamprecht, I., and A. I. Zotin, eds. *Thermodynamics and the Regulation of Biological Processes.* Berlin: Walter deGruyter.

——. 1985a. *Anticipatory Systems: Philosophical, Mathematical and Methodological Foundations.* New York: Pergamon Press.

——. 1985b. Information and cause. Presented at 20th Annual Orbis Scientiae, University of Miami, Jan. 17–21. In Mintz, S. L., and A. Perlmutter, eds. *Information Processing in Biological Systems,* 31–54. New York: Plenum Press.

——. 1985c. The physics of complexity. Second W. Ross Ashby Memorial Lecture, at 7th European Meeting on Cybernetics and Systems Research, Vienna, Austria, April 27. *Systems Research* 2: 171–5.

——. 1985d. Organisms as causal systems which are not mechanisms. In Rosen, R., ed. *Theoretical Biology and Complexity*, vol. VI, 165–203. New York: Academic Press.

——. 1987a. On the scope of syntactics in mathematics and science: The machine metaphor. Proceedings of Workshop on Brain Research, Artificial Intelligence and Cognitive Science, Abisko, Sweden, May 12–16, 1986. In Casti, J. L., and A. Karlqvist, eds. *Real Brains, Artificial Minds*, 1–23. New York: Elsevier Science.

——. 1987b. Some epistemological issues in physics and biology. In Hiley, B. J., and F. D. Peat, eds. *Quantum Implications: Essays in Honor of David Bohm*, 327–41. London: Routledge, Kegan and Paul.

——. 1988a. Effective processes and natural law. In Herken, R., ed. *The Universal Turing Machine: A Half-Century Survey*, 523–37. Oxford: Oxford University Press.

——. 1988b. How universal is a universal unfolding? *Applied Mathematics Letters* 1(2): 105–7.

——. 1988c. System closure and dynamical degeneracy. *Mathematical and Computational Modelling* 10(8): 555–61.

——. 1989. The roles of necessity in biology. Proceedings of 1987 Abisko Workshop on Process, Function, and Form, Abisko, Sweden, May 4–8. In Casti, J. L., and A. Karlqvist, eds. *Newton to Aristotle: Toward a Theory of Models for Living Systems*, 11–37. Boston: Birkhauser.

——. 1991. *Life Itself: A Comprehensive Inquiry into the Nature, Origin, and Fabrication of Life*. New York: Columbia University Press.

——. 1993. Drawing the boundary between subject and object: Comments on the mind-brain problem. *Theoretical Medicine* 14: 89–100.

——. 1994. Mind as phenotype. Lecture given at Karolinska Institute, Stockholm.

Rosenblatt, F. 1962. *Principles of Neurodynamics, Perceptions and the Theory of Brain Mechanisms*. Washington, D.C. Spartan Books.

Savageau, M. 1976. *Biochemical Systems Analysis*. Reading, Mass.: Addison-Wesley.

Schrödinger, E. 1944. *What Is Life?* Cambridge: Cambridge University Press.

Shannon, C., and J. McCarthy, eds. 1956. *Automata Studies*. Princeton, N.J.: Princeton University Press.

Shannon, C. E., and W. Weaver. 1949. *The Mathematical Theory of Communication*. Urbana, Ill. University of Illinois Press.

Sundermeyer, K. 1982. Constrained dynamics. In *Lecture Notes in Physics*, vol. 169. Berlin: Springer-Verlag.

Swan, G. W. 1984. *Applications of Optimal Control Theory in Biomedicine.* New York: Dekker.

Thom, R. 1976. *Structural Stability and Morphogenesis.* (Trans. D. H. Fowler.) Reading, Mass.: W. A. Benjamin.

Thompson, D'Arcy W. 1917. *On Growth and Form.* Cambridge: Cambridge University Press.

Turing, A. M. 1950. Can a machine think? *Mind* 59: 236.

von Bertalanffy, L. 1932. *Problems of Life: An Evaluation of Modern Biological and Scientific Thought.* London: Watts. Reprinted 1952, New York: Harper.

von Bertalanffy, L., W. Beier, and R. Laue. 1977. *Biophysik des Fliessgleichgewichts.* Berlin: Akademie-Verlag.

von Neumann, J. 1955. *Mathematical Foundations of Quantum Mechanics.* Princeton, N.J.: Princeton University Press. [German ed., 1932.]

———. 1966. *Theory of Self-reproducing Automata.* (Assembled by A. W. Burks.) (See lecture 4.) Urbana, Ill. University of Illinois Press.

von Neumann, J., and O. Morgenstern. 1953. *Theory of Games and Economic Behavior.* Princeton, N.J.: Princeton University Press.

Waddington, C. H. 1950. *Biochemistry and Morphogenesis.* Cambridge: Cambridge University Press.

West, B. J. 1990. *Fractal Physiology and Chaos in Medicine.* Singapore: World Scientific.

Wheeler, J., and W. H. Aurek. 1983. *Quantum Theory and Measurement.* Princeton, N.J.: Princeton University Press.

Wiener, N. 1961. *Cybernetics.* (See chapter 11.) (1st ed., 1948.) New York: MIT Press.

Wigner, E. 1960. On the unreasonable effectiveness of mathematics. *Communications in Pure and Applied Mathematics* 13: 1–14.

Index